Praise for *The Einstein Effect*

"A fascinating and funny guide to history's favorite genius—and why he still matters."

—A. J. Jacobs, *New York Times* bestselling author

"In creating the appearance of Doc Brown in *Back to the Future*, Albert Einstein was my obvious inspiration because, to portray a brilliant physicist, I wanted to look like a brilliant physicist! I hope this book encourages others to be inspired by Einstein as well."

—Christopher Lloyd, award-winning actor

"Albert Einstein left us some seven decades ago, but the world he created lives on—and there's no better tour guide to it than Benyamin Cohen. He serves up the science, of course, but thankfully lightened with a healthy dose of pop culture and humor. A book for geniuses and the rest of us alike."

—Derek Baxter, author of *In Pursuit of Jefferson*

"You don't have to be a distant relative of Albert Einstein like yours truly to appreciate Albert's profound and enduring effect on the world around us. All you need to do is read Benyamin Cohen's thoroughly entertaining new book. A full head of hair might help also."

—Jeff Einstein, great-great nephew of Albert Einstein

"An extraordinary, whirlwind adventure into all things Einstein, examining why the life and work of the 20th century genius still inspires and impresses us today. From Einstein's actual brain to the modern technologies that his theories anticipated, Cohen offers a masterful and astounding tour. Full of humor and surprises, *The Einstein Effect* is the perfect read for fans of a larger-than-life figure whose heartfelt dedication to human rights rivaled only his groundbreaking science."

—Paul Halpern, physics professor and author
of *Einstein's Dice and Schrödinger's Cat*

Also by Benyamin Cohen

*My Jesus Year: A Rabbi's Son Wanders
the Bible Belt in Search of His Own Faith*

THE EINSTEIN EFFECT

How the World's Favorite Genius Got into Our Cars, Our Bathrooms, and Our Minds

BENYAMIN COHEN

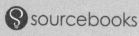

Published by Sourcebooks
P.O. Box 4410, Naperville, Illinois 60567–4410
(630) 961-3900
sourcebooks.com

Cataloging-in-Publication Data is on file with the Library of Congress.

Printed and bound in Canada.
MBP 10 9 8 7 6 5 4 3 2 1

To Albert Einstein and all of those who he described as "passionately curious"

CONTENTS

EINSTEIN TWEETS

"If the facts don't fit the theory, change
the facts. —Albert Einstein."

—IVANKA TRUMP

W hen Ivanka Trump shared this famous Einstein quote with her six million Twitter followers after lunch on a sunny summer afternoon, she wasn't hoping to make news. It was—like so much of the ones and zeros that zoom across the digital ether—just something to share on a Wednesday. It was an innocent post, but there was just one problem…

Albert Einstein never actually said that.

And the irony wasn't to be missed. She was tweeting something that was factually inaccurate in a tweet about changing the facts. Fortunately for Ivanka, the internet is flooded with incorrect Einstein memes, so she could be forgiven for finding one of the many misquotes that reside across the web. Indeed, nobody even noticed the tweet, let alone the mistake.

That is until Albert Einstein himself rose from the dead to correct her.

At 8:15 a.m. on July 24, 2017, the verified Albert Einstein

Twitter account sent out a message to Ivanka. He confirmed that he never said that and then, for good measure, included a link to a book of official Einstein quotes. It was the mic drop heard round the world. It made international news. The *Huffington Post*'s headline: "Ivanka Trump's attempt to quote Albert Einstein backfires spectacularly." *Newsweek* declared: "Ivanka Trump misquoted Einstein and the Internet loves it."

In an age of "fake news," this was seen as a triumph.

I know this story firsthand because from my home nestled in the forests of the Appalachian Mountains, I log onto the internet each morning and manage the official social media accounts of our planet's favorite genius. After collecting eggs from my chickens and helping my neighbor milk her cows, I boot up my computer and pretend to be a Nobel laureate. To millions of people around the world, I am the digital avatar of Einstein (my short and kempt hair not withstanding). Teenagers in India message me for help with their science homework, physicists in Florida email me the findings of their latest research, producers at PBS call and ask if I'll promote a new Einstein documentary. It's an awesome responsibility, one which I don't take lightly, because I know that Einstein's contributions to our understanding of science are at least matched, if not surpassed, by his importance as a symbol.

That's because Albert Einstein is widely considered to be the first modern-day celebrity. Before Jennifer Lopez and Ben Affleck, paparazzi were chasing around the world's favorite scientist. They wanted to know the most mundane details about his life: How many hours did he sleep each night? What did he eat for breakfast? "Pressmen roared up the plank whenever his ship docked in America," wrote Carolyn Abraham, who published a history of Einstein. "A couple of them once fell into the harbor trying to scramble on board." That iconic photo of Einstein sticking out his tongue? That was his way of acknowledging the ludicrousness of his situation: He was a German Jewish refugee seeking asylum from Hitler's wrath. He was famous for his high IQ, for physics and math and quantum mechanics—and yet, here he was in Los Angeles visiting movie studios and tossing bon mots at a red-carpet Hollywood premiere with his friend Charlie Chaplin.

Nearly seven decades after Einstein's death, he is still a celebrity. Nowhere is this more evident than on social media, where he has verified blue-checked accounts, with nearly twenty million followers on Facebook (more than Tom Hanks!), another million on Instagram, and more than half a million on Twitter. And I get to talk to all those people each day as the wizard behind the curtain:

I am Albert Einstein.

Dead celebrities are surprisingly popular on social media. On Facebook, you can be friends with Elvis Presley or John Lennon—both of whom each have about thirteen million fans. Follow Marilyn Monroe on Instagram (1.8 million followers) for fashion inspiration. John Wayne is on Twitter, but he doesn't have much to say. That may explain why he has fewer than forty thousand followers. Yet Einstein dwarves them all in both size (did I mention the more than twenty million fans?) and content: I share roughly ten posts a day. There's Throwback Thursday photos of a young Einstein posing with his sister and the famous quotes I post for Wednesday Wisdom. There's also the video I made of an Einstein bobblehead sledding down a hill in a snowstorm and the countless contemporary science news headlines that are surprisingly related to Einstein's research.

Most of these accounts are run by the estates of the dead celebrities, and my case is no different. I work for the Albert Einstein Archives. When Einstein died, he bequeathed his estate to the Hebrew University in Jerusalem (more about that in a later chapter). The archivists and curators who work there are all academics; they mostly help find material for other academics doing research about Einstein. Tweeting is not really their expertise. So that's where I come into the picture. I'm the hired hand who can, on a good day, come up with a clever retort to Ivanka Trump.

Keep in mind, I'm not a scientist. At best, I'm a likable idiot. The Einstein family maid once dubbed Albert "the dope," so maybe there's hope for me. But I think there's a benefit to not knowing the difference between a quantum mechanic and an auto mechanic. The majority of people who are following Einstein online likely couldn't explain the theory of relativity any better than I could, yet they still feel a connection to this genius. This is nothing new: When Einstein gave a speech to a packed audience at the Kroll Opera House in Berlin in 1930, the headline in the newspaper the next day was "4,000 Bewildered as Einstein Speaks." Sure, he represented the romance of the cosmos, but he transcended science.

Wharton Professor Jonah Berger, author of *Contagious: Why Things Catch On,* spent years researching the most shared stories on the *New York Times* website. He found that articles about science and innovation tend to go viral because they evoke a sense of awe. And our brains are wired—and we're inherently inclined—to want to share stories of discovery. Awe-inspiring articles were 30 percent more likely to make the *Times'* "Most E-mailed" list. As Albert Einstein himself noted, "The most beautiful emotion we can experience is the mysterious. It is the power of all true art and science. He to whom this emotion is a stranger, who can no longer pause to wonder and stand rapt in awe, is as good as dead." It's this aspirational "aha" attitude that helps make Einstein—more than, say, John Wayne—so popular online.

In my role as Einstein, I don't don a wig and glasses, but rather a digital skin, an embodiment of genius to the masses. When the guy in Great Britain tweets at Albert, little does he know that a short, bespectacled Jewish guy in rural West Virginia is pretending to be perhaps the greatest, and definitely the most famous, scientist who ever lived.

Managing Einstein's social media accounts is more than just posting inspirational quotes and archival photos. People also message asking me to explain gravitational waves, lidar (light detection and ranging), and qubits. I kid you not, while writing that sentence, I had to look up how to spell *lidar* and *qubits*. But I have experience being an impostor. More than a decade ago,

I wrote a memoir called *My Jesus Year: A Rabbi's Son Wanders the Bible Belt in Search of His Own Faith* in which I took off my skullcap and crossed the proverbial street to another faith. A life-long synagogue-goer (indeed, we had a synagogue attached to our home), I visited fifty-two churches in fifty-two Sundays and discovered something remarkable: hanging out with Jesus actually made me a better Jew. So, when I got the opportunity to "be" one of the most famous people in history, I jumped at the chance to tackle another persona.

What I've found along the way is that Albert is more than just a chapter in history. His work lives on in so many facets of our everyday lives—in science, technology, entertainment, and so much more. I merely serve as the tour guide to a planet full of people who look to Einstein as their muse.

When I first took the job of becoming Albert, I thought I knew enough about the man to skate by. I could write a list of the ten most iconic mustaches of the twentieth century. (Here's looking at you, Mark Twain.) What's more, since I had listened to the entire twenty-one-hour audiobook of Walter Isaacson's seminal Einstein biography, I was surely qualified to shoot off some tweets. But what I soon discovered was that Einstein is not someone who merely exists in the past. Unlike other great scientists of old—like Isaac Newton and Thomas Edison—whom we revere but whose discoveries we have largely moved on from and improved upon, Einstein's theories and equations are still as alive and essential as they ever were.

Decades after his death, Einstein is still making the mundane magical. So much of the technology we touch on a daily basis—and that shapes how humans see the universe is a direct result of Einstein's genius. This technology may be invisible to the naked eye, but take a closer look: His work lives on in the form of iPhone cameras and burglar alarms, remote controls, supermarket scanners, laser eye surgery, and in the space program. Driverless cars, DVD players, weather forecasting, and even the search for aliens—it's all thanks to theories hatched by Einstein. His formula to measure the size of molecules dissolved in liquids made it possible to create or improve thousands of consumer products—including better shaving creams and toothpastes. Not to mention the more than fifty inventions he had patented while he was still alive—including an eco-friendly refrigerator that required no electricity. His work on relativity, gravity, and curved space-time enabled the invention of GPS devices, which were first created to guide satellites in outer space—and are now used to tell you when your pizza delivery will arrive. (What has Galileo done for you lately?)

But it's not just his scientific discoveries that continue to shape our world. In 2023, more people know Einstein as an icon than as a theorist. Googling "Einstein icon" nets you millions of results—including actual emoji of Einstein's iconic face. He's the one scientist that everyone knows on sight. Walk up to any elementary school student and ask them to name a genius: it'll likely be Einstein.

———

"Do not believe everything you read on the Internet,
especially quotes from famous people."

—ABRAHAM LINCOLN

Becoming Albert Einstein has led me to ask some larger questions: Is there a place to celebrate genius in a post-literate world filled with scrolling news feeds of cat videos and Instagram posts of influencers on vacation (and photos of Lincoln taking a selfie)? In an age when ignorance is often celebrated, what do Einstein and his "big brain" stand for? His accessibility as a genius—not to mention his own popularity on social media—gives Einstein entrée and relevance to an entirely new generation. That's the point of giving him an online presence long after his death: to keep his spirit of discovery and drive for knowledge alive. As the voice of one of the planet's most respected minds, I sit at the crossroads of "Einstein the celebrity" and "Einstein the scientist." My job is to continue to build Einstein's popularity, but not just for celebrity's sake; the articles I share with millions each day cover cutting-edge science and new discoveries, many of which are directly related to Einstein's own work. But there is so much more of the story to tell.

In his 2018 book *The Death of Expertise*, Professor Tom Nichols wrote how in our age of unlimited knowledge—where answers are just a Google search away—the internet is actually making all of us a lot dumber. Sure, we no longer need to memorize the numerical value of pi or the capital of North Dakota; Google's vast servers can store that information for us. (It's 3.14 and Bismarck…in case you were wondering.) But that shift in our cultural mindset has produced a downside as well—many, in fact. That glut of readily available information has created the feeling that we have less of a need for actual experts, not to mention an increase in those who like to call themselves experts. For example, asked during the run-up to the 2016 presidential election whom he turns to for advice on foreign affairs, Donald Trump famously cited himself.

The internet has given us all free Gmail accounts, but it has also birthed the likes of 4Chan and Reddit, online message boards that are littered with the detritus of misinformation, conspiracy theories, and bogus science. Anti-vaxxers and flat-earthers can now easily find comrades in their own self-created filter bubbles, magnifying and emboldening their views. At the same time, we've seen an alarming increased notion of anti-elitism and anti-intellectualism. Career State Department officials, who have spent decades in the trenches, are now no smarter than a troll with a Twitter account. Pulitzer Prize–winning journalists are being accused of writing fake news.

Coalesce all of these disparate threads and we can ask ourselves a proverbial question. I call it WWED: What Would Einstein Do? What would Einstein think of all this, especially when it comes to the universal search for knowledge? How would he react to a 2020 law passed in Ohio called the Student Religious Liberties Act that allows students to give incorrect answers on a test if giving the correct answer is against their religion? (Put simply, a student could write a wrong answer about evolution or climate change and still ace the test.)

Einstein was deeply philosophical and famously open to

questioning our assumed presumptions about the world. But when it came to facts, he spent a lifetime trying to prove his work. He would likely not take too kindly to those sowing doubt and spewing misinformation. He was not one to stay silent and keep his thoughts to himself. He would be the first to mock the anti-vaxxers with a pithy tweet. Indeed, he reportedly once told a friend, "If you think intelligence is dangerous, try ignorance."

That's why I've decided to go on this immersive quest to find the modern-day relevance of Einstein, my own "unified field theory" of Albert. This journey will take me across continents: from the United States to Israel and Japan. I'll stand atop the world's largest telescope, become an Uber driver, and meet a man who built a time machine—all to show how Einstein's theories are weaving their way into our daily lives. I will interview dozens of Einsteins (including not one, but two Rabbi Einsteins) and spend months tracking down the last remaining physical vestige of Einstein—his brain, stolen from his body by the pathologist performing Albert's autopsy. (Spoiler alert: it's kind of squishy.)

In an era when we are flooded with information (and misinformation) and we don't know whom to believe, the celebrated mind of Einstein is needed more than ever. He is someone we can all believe in. But, even beyond the man himself, our landscape is replete with modern incarnations of Einstein—scientists, artists, and those in a whole host of other fields who are inspired by the greatest mind of the twentieth century. In their own ways, they are each continuing Einstein's legacy—not just with their work, but also by becoming new sources of inspiration for the rest of us. Like Einstein, they look at the world with a sense of curiosity and a desire to ask questions. Those key ingredients are often what's missing in our national dialogue.

Einstein is not lost to the annals of history, only to be read about in dusty classroom books, or plastered with a quote on coffee mugs. His theories and his worldview are as alive today as they ever were. And his quest for deeper understanding should be a call to arms in our age of short TikTok videos.

Managing the official social media accounts of Albert Einstein is not your typical gig. He is a larger-than-life character. Indeed, he was *Time* magazine's Person of the Century—ahead of the likes of MLK, FDR, the Beatles, Princess Diana, and Gandhi. Einstein would want that popularity put to good use to make us all scientists of some kind or another.

My role is as an Einstein evangelist, and I preach his gospel every day. My goal is to infuse others with that same curiosity. Many already look to him as a symbol of genius, but they aren't always connecting that with how vital Einstein's spirit of inquiry and dedication to science are to our everyday world. I want readers to experience that same sense of wonder so many budding scientists and veteran experts have for Einstein, with the hope that it inspires them to be more vigilant and questioning with the information they find online.

There have been more than seven hundred books written about Einstein. "It's like trying to drink from a fire hose," Dr. Fred Lepore, who wrote one of those books, told me. Why do we need one more? Sure, I get the question. But isn't it the same as asking the executives at CBS why we need yet another cop drama? Each one, in its own unique way, adds something to the canon.

What follows is my own quixotic quest to discover Einstein's invisible hand impacting the modern world and showcase the wide variety of people who are inspired by him as they make the discoveries and advances of tomorrow.

STEALING EINSTEIN'S BRAIN

"We should take care not to make intellect our god; it has, of course, powerful muscles, but no personality."

—ALBERT EINSTEIN

The day I held Albert Einstein's actual brain, I threw up. To be clear, I didn't heave from seeing the brain. (We'll get to why I threw up a little later in this story.) Indeed, the brain itself is, in some ways, a thing of beauty. Sure, it's been chopped up into more than two hundred pieces, wrapped in gauze, immersed in alcohol for preservation purposes, and stored in several mason jars. Some people use mason jars to drink artisanal cocktails. Others, apparently, use it to store the greatest mind of the twentieth century.

Einstein looked like everyone's beloved grandfather. But his brain looked like stale pieces of cauliflower. Some bits were just floating freely, bobbing around much the same way Einstein himself did while sailing, one of his favorite pastimes. This organ had spent seventy-six years inside the cranium of perhaps the most important scientist who ever lived. The cerebellum that gave us $E=mc^2$ was right here in front of me. Perhaps it held the key to wisdom, to genius, to life itself.

These jars containing the last remaining physical vestige of Albert Einstein were not, as you might imagine, kept behind lock and key. Nor are they stored in a climate-controlled exhibit at a fancy museum or in the archives of a prestigious university. The day I saw the brain, it was in the back of an Acura MDX. So, there I was, with my head inside the trunk of a mid-market SUV, peeking at history. Someone driving by could've easily thought I was buying stolen electronics from a low-level mobster.

The mason jars were, thankfully, not just rolling around in the trunk alongside groceries and jumper cables. I mean, the guy driving around with Einstein in the back of his car is not a Neanderthal. The jars were placed in a decades-old cardboard box, stamped with the logo of the Kaweah Citrus Association. It used to hold oranges. Now it contained Einstein's brain. Across the side of it were the numbers "55–33," scribbled there in marker to denote the year 1955, when Albert Einstein, upon his death that April, became the thirty-third autopsy performed at Princeton Hospital.

And that's where this sordid tale begins. The doctor who conducted Einstein's autopsy stole the brain. And here I was, decades later, staring right at it.

"The last words of the intellectual giant were lost to the world," the *New York Times* wrote when it broke the news of Einstein's passing. "The only person at his deathbed, Mrs. Alberta Rozsel, the night nurse, said he mumbled in his sleep several words in German that she did not understand." I'd like to think he was telling her the key to the inner workings of the universe, now forever lost to history because Rozsel had failed to learn a second language in high school.

Albert Einstein died at 1:15 a.m. on April 18, 1955. Dr. Thomas Harvey, the pathologist on duty that day at Princeton Hospital, began his autopsy shortly after dawn. He took out different organs for examination and, as was routine at the end of such a procedure, Harvey went for the brain. Literally. He took an electric saw to Einstein's scalp. If you were walking down the hallway, it would've sounded like someone was cutting down a Douglas fir in the next room. Or sawing a man in half. Harvey would also use a mallet, tweezers, and his own fingers to eventually wrest the brain from the cavity of its wise owner. "To me, it was obvious that the brain of this man should be studied," Harvey would later recall. "It was the biggest moment of my life."

At 11:15 a.m. on that warm spring morning, there was a press conference on the hospital steps. Dr. Harvey, dressed in a crumpled lab coat with a pen in its front pocket, told the crowd of reporters that the death had been caused by an abdominal aortic aneurysm, describing it as "a big blister on the aorta which broke finally like a worn-out inner tube." In other words, an aneurysm had burst in Einstein's stomach, causing internal bleeding. He had first complained of belly issues years earlier, in 1948. Doctors discovered a grapefruit-sized bulge in his abdomen, treated it as best they could, and warned Einstein that it would eventually burst without invasive surgery.

"Let it burst," Einstein told them, and seven years later, it did just that.

While Harvey was outside speaking to the press, Einstein's ophthalmologist, Henry Abrams, rushed downstairs to grab the eyeballs from his famous patient, as if it was an anatomical yard sale. *All items must go!* Abrams did, in fact, abscond with the eyes, which now, all these years later, are rolling around in a bank safety deposit box in New Jersey. At one point, Abrams was apparently offered $5 million for them—by none other than pop star Michael Jackson. But I digress.

At 2:00 p.m., Einstein's body was taken from the morgue to the nearby Ewing Crematorium in Trenton. As per the scientist's wishes, his body was burned, the ashes scattered in an undisclosed location. The last thing Einstein wanted was for his grave to become a celebrity tourist attraction. His life's work would be his legacy. In death, he wanted some peace from the adoring crowds. That peace, however, would not last long.

Two days after Einstein's death, Harvey announced that he would assemble "a team of outstanding medical men" to study the brain at a conference the following week. Harvey wanted to find out if there was something special about Einstein's brain that made him so smart. But Otto Nathan—who managed Einstein's finances, was a trusted friend, and served as the executor of his estate—was furious when he found out what Harvey had done. He did not want the brain of his beloved client and friend to become an object of fascination, the very thing Einstein would've despised. Nathan wanted, as he wrote to Harvey, "the wishes of the deceased and family be properly respected." Harvey had no money to set up such a conference and the ghastly exercise never took place.

Meanwhile, Harvey photographed the brain from every angle and then, for the sake of science, chopped it up, dissecting it into 240 blocks. He further sliced some of the parts into wafer-thin pieces to create one thousand microscopic slides of the brain tissue for further study. The slides were then put into ten small pine boxes. There were still plenty of blocks left and the remainder of those went into mason jars filled with alcohol to help with preservation. The jars started off in Harvey's Princeton office, then he moved them home where they rested on the mantel in the living room, and eventually ended up in a box in his basement.

Harvey's harebrained idea, to see if he could ascertain the basis of Einstein's genius by examining the actual brain tissue, was not entirely insane. In the early 1900s, anatomists were already ascribing musical genius to the unique qualities of the brains of classical composers Beethoven and Bach. In 1929, a neurologist named Oskar Vogt was hired to study the brain of Vladimir Lenin. After slicing it into 30,953 wafers and pickling it for preservation, Vogt claimed to have discovered the source of Lenin's political smarts: many unusually sized "giant cells" that had connected parts of the brain that were separate in the average person. Vogt said this proved Lenin's genius and was the reason for the "multiplicity of his ideas" and his "powers of intuition." Just ten years before Einstein's death, in 1945, a U.S Army doctor had studied the brain (or whatever was left of it) from the gunned down skull of Benito Mussolini. In the 1990s, someone decided to preserve the brain of Jeffrey Dahmer, the notorious serial killer.

Studying brains was nothing new. A group of doctors created the American Anthropometric Society in 1889 in Philadelphia. Whenever someone in the group died, their brain would be donated for the others to study. Walt Whitman was apparently so enamored with the group that he volunteered his own brain. So, when he died in 1892, an autopsy was performed by Henry Ware Cattell, who removed Whitman's brain, preserved it in a jar, and handed it over to a bumbling assistant who dropped it, destroying any physical remnants of America's favorite poet. The incident is said to have inspired the 1931 *Frankenstein* movie starring Boris Karloff in which an assistant drops a container labeled "normal brain" and instead gives Dr. Frankenstein one labeled "abnormal."

But brains are not the only skeletal souvenirs people keep. The sister of composer Frédéric Chopin kept his heart in a crystal jar and pickled it in Cognac. Grave robbers stole the corpse of Charlie Chaplin, a good friend of Einstein's, and held it for ransom. Mobsters attempted to steal President Lincoln's body from its tomb. Even Harvey with his handsaw was not the first pathologist to steal a famous body part.

On May 5, 1821, in an autopsy suite filled with dozens of spectators, the pathologist examining the fifty-one-year-old Napoleon Bonaparte pilfered the penis. An article in the *Journal of Sex Research*, alliteratively titled "The Peripatetic Posthumous Peregrination of Napoleon's Penis," expounds on the postmortem life of the appendage. It's been passed through many hands since it was freed from its human body two centuries ago and was, at one point, on display in New York at the Museum of French Art. A visiting tourist could see art from Renoir, Degas, and Bonaparte's penis. Napoleon's little Napoleon—described through the years as both a "mummified tendon" and a "shriveled eel"—now belongs to the daughter of a urologist.

Thomas Harvey and Princeton Hospital eventually split ways, with Harvey taking the brain with him on his way out.

Over the years he stored it in various places: in a beer cooler, under a pair of socks, and in Tupperware containers. Every so often, he'd contact scientists in the field, inquiring if they wanted to examine a section of Einstein's gray matter. Pieces of the brain have since been found in Tokyo, Buenos Aires, and Honolulu. They were often sent in unassuming brown envelopes via the U.S. Postal Service. Harvey once sent a few pieces inside a FedEx envelope to a professor at Rutgers University where a student placed it in a jar labeled "Big Al's Brain." He also shipped four pieces in a mayonnaise jar, and one time, it got lost in the mail on its way to Alabama.

In the years that followed, most people forgot about Harvey and the world's most famous cerebral cortex. Occasionally, Harvey would show it off at his local Rotary club. But besides a brief article about the brain that appeared days after Einstein's death, it seemed like it had not been reported on for two decades. That is, until 1978, when journalist Steven Levy got the strangest assignment of his young career. "I want you to find Einstein's brain," his editor told him. Levy eventually tracked down Harvey, who was living in Kansas at the time, and flew out to meet with him. Harvey was at first reluctant; like Einstein, he was a reticent celebrity. But eventually Harvey dug into a cardboard box, grabbed a few jars, and handed them to Levy. "For a moment, with the brain before me, I had been granted a rare peek into an organic crystal ball," Levy wrote in his story, "My Search for Einstein's Brain," which was published in that August's edition of *New Jersey Monthly*. "Swirling in formaldehyde was the power of the smashed atom, the mystery of the universe's black holes, the utter miracle of human achievement… It is something of ourselves at our best."

Levy's article brought the story to the modern masses. Newspapers from across the globe wrote about it. Johnny Carson cracked jokes about it. "*Good Morning America* called and *The Today Show* and all sorts of others, from Japan and from Australia," said Cheryl Schimmel, who managed Harvey's office. "They all offered him good money to come on TV and talk about it, but he wouldn't even take their calls. He just wasn't interested in things like fame or money." Levy had written that the small pieces of the chopped-up brain reminded him of Goldenberg's Peanut Chews. Without missing a beat, the candy company sent the journalist a giant box to thank him for the free publicity.

A subsequent 1994 story about the brain in the *Wall Street Journal* caught the attention of Gregory Stock, an ambitious biotech entrepreneur. He thought he could help Harvey make money by selling slices to museums and individual collectors for millions of dollars. "It reminded me of Jesus Christ and the pieces of the cross," Stock said. "Fake chips of them had been sold and resold through the ages for fortunes, and here was the real thing: the brain of Albert Einstein. Nothing could equal it. And if science was the modern surrogate for religion, this was a holy relic." Harvey, down on his luck, really could've used the money. But he ultimately turned down Stock's offer. "It was just too psychologically difficult to give up," Stock said. "It was his connection to something larger than himself. It had become the meaning of his life."

In 1995, a group of rabbis wrote to Harvey asking if they could give the brain a proper Jewish burial. Harvey, a lifelong Quaker, wasn't ready to part with it just yet. "I thought this would be okay when all the research was finished, which it wasn't, I didn't think," Harvey said.

As Harvey got older, he began to feel the weight of his fateful decision to steal the brain. In 1997, Michael Paterniti, a freelance magazine journalist, approached an eighty-four-year-old Harvey with a proposition: Why not return pieces of the brain to Einstein's granddaughter, Evelyn, who lived in California? Paterniti offered to drive Harvey across the country in one of the most unusual road trips I can imagine. In a rented Buick Skylark, Paterniti and Harvey sat up front, with Einstein's brain bobbing around in a Tupperware container in the trunk. When they arrived to see Evelyn, Harvey changed his mind and returned home without ever giving her a piece. "I could have had it as an amulet for a necklace, or a curiosity to take out to show people when one is feeling particularly ghoulish," she later told an interviewer.

Paterniti turned the quixotic adventure into a book, *Driving*

Mr. Albert, which I read as I began my own career in journalism. His book had a huge impact on me, and, in a sense, it was my gateway drug. That's because my entire fascination with all things Einstein began the moment I discovered that someone had stolen the physicist's brain. Like most people who grew up learning about Einstein, I had heard about (if not fully understood) the theory of relativity, the Nobel Prize, the wild hair. But not one teacher ever taught us about the doctor on duty that April morning when Einstein died.

Perhaps nobody alive today knows more about Einstein's medulla oblongata than science journalist Carolyn Abraham, who wrote a book, *Possessing Genius*, about the journey Einstein's brain took from the autopsy table to the myriad people who touched it and studied it over the next several decades. She spent three years of her life researching the story, had interviewed Harvey and gained his trust. I reached out to her and peppered her with questions. How did she first find out about the heist? What was Harvey like in person? What was her reaction when she saw the brain? We talked for hours.

As our conversations continued, Carolyn told me she was working with a Toronto production company to turn her book into a documentary for the Canadian Broadcasting Corporation. They would be traveling to the United States to interview Harvey's children on camera, and she wanted to know if she could come to West Virginia to visit me. I was confused. What did I have to do with the story?

"You have a unique take on Einstein," she replied. She explained that the movie would have two elements: What happened to the physical pieces of the brain, but also the legacy that the brain continues to inspire today. "That's where you come in," she said, citing the work I was doing sharing Einstein news to his millions of followers on social media. It didn't hurt that I lived in the woods and had a flock of chickens—including one hen with a bouffant of white feathers we named Alberta Einstein. "You'll be the comic relief for an otherwise macabre story."

And so that's how Carolyn and a Canadian film crew ended up at my home. As the producer and camera crew lugged equipment out of their van, Carolyn approached my pugs, Fergus and Spike, with the same enthusiasm she did the brain: curious and playful. She was short and tan, her glasses propping up her shoulder-length auburn hair. As lights and tripods were set up in my office, Carolyn and I sat around my kitchen island, where I had set up a spread of bagels, cream cheese, and coffee for the guests. I asked her to tell me again how she first heard about the brain.

She began her career as a crime and politics reporter. Science journalism was the furthest thing from her mind. "At the end of all my science and math in high school, I think I had actually done cartwheels," she told me. But her editors saw something different: a quirky sensibility. This was in the late 1990s, at the dawn of a new millennium, when scientific discoveries were spilling over into the mainstream. The human genome was on the precipice of being mapped, the first embryonic stem cells were about to be harvested, and a sheep named Dolly had just been cloned. "Science was raising a lot of interesting social and ethical questions that made the whole field more interesting to me," Carolyn said, as she took a bite from her everything bagel.

She was still new at the job when, in 1999, she got a tip about a Canadian neuroscientist, Sandra Witelson, who had received a piece of Einstein's brain to study. Carolyn ended up writing an article for the *Globe and Mail* about Witelson and Harvey, who had rumbled across the border with Einstein's brain in the trunk. Harvey told Carolyn that he pulled over as he entered Canada. "I thought I'd better stop at customs and tell them I had this brain."

Witelson was well regarded; her laboratory at McMaster University had one of the largest collections of "cognitively normal" brains in the world. What's more, she had information about each of the people to whom those brains belonged. She used that data to study things like the differences between the brains of men and women, and the brains of heterosexuals versus homosexuals. While examining Einstein's brain, Witelson found something remarkable: his parietal lobes were 15 percent wider than a control group of ninety-one other brains. "This unusual brain anatomy may explain why Einstein thought the way he did," Witelson said. "Einstein's own description of his scientific thinking was that words did not seem to play a role. Instead, he saw, more or less, clear images of a visual kind." Could this have made it possible for Einstein to visualize a concept like relativity?

Witelson published her findings in an article titled "The Exceptional Brain of Albert Einstein" in *The Lancet* medical journal and, as expected, it made international news. One newspaper headline jokingly declared: "Einstein was bigger where it counts." Witelson herself became something of a celebrity, giving a TED Talk and appearing in *People* magazine. McMaster University gave her $1 million to do more research and honored her with the first-ever Albert Einstein–Irving Zucker Chair in Neuroscience.

"The timing of that paper was notable because it was the end of the century," explained Carolyn. "Everybody was making lists: Who was the top scientist of the twentieth century? What were the greatest advances in science and who made them? Einstein was topical in 1999. There was a certain confidence in all things scientific. There was this sense of what can't we do?"

For Thomas Harvey, it was, finally, vindication. "He kind of started to regard himself as having been this time capsule," Carolyn said, "just transporting the brain into the future to a time when science might figure this out."

Harvey died from a stroke in 2007 at the age of ninety-four. His family discovered the remaining slides containing slivers of Einstein's brain while cleaning up the house where he had been living and decided to donate them to the National Museum of Health and Medicine, located on a military base in Silver Spring, Maryland. Government officials showed up at the Harvey residence and loaded everything into the back of a van and drove off. At one point, realizing the trove they now had, the museum digitized the slides for public consumption. They built an iPad app, and for $9.99, anyone could download the microscopic remnants of Einstein's brain. "I'd like to think Einstein would have been excited," said Steve Landers, who designed the app. (The program has since been removed from Apple's app store and images of the slides have been scrubbed from the museum's website for reasons unknown.)

When the museum officials were in Harvey's basement, they discovered something else: a cache of never-before-seen photos that Harvey had taken of the brain before he sectioned it. For researchers, these pictures of Einstein's brain in its entirety are priceless. But this material is not open to the public and is somewhat shrouded in mystery. It resides, locked up, on a military base. One of the few researchers given access was Fred Lepore, a neurology professor at Rutgers University. Lepore and his colleague, Dr. Dean Falk, an anthropologist at Florida State University, spent eight hours inside the facility poring over the photographs of the full brain.

For Lepore, who lives blocks away from Einstein's former home on Mercer Street in Princeton, seeing Einstein's brain was not exactly a holy experience. "I didn't hear an ethereal choir in the background," he told me when we chatted via video call. But it did change him. "These are the mortal coil, the trappings of one of the most exceptional people to ever grace the Earth."

The first thing Lepore noticed was that it was not a particularly large brain. "It was 1,230 grams that saw the universe like nobody else," he said. He was particularly interested in this question: Could the physical contours of a brain explain genius? There are two sides to this debate: On one side, you have neurologists like Lepore who see the brain as the body's engine. "If there's something wrong with the right side of your brain, I know that the left side of your body is not going to work so well. But a lot of people say, 'No, you can't reduce the soul, spirit, consciousness, and genius to three pounds of gray glob inside your skull.'" To this day, Lepore said, experts are unsure how "a little piece of gray matter generates an idea or concept. We have no idea. It's an incredibly profound question."

What Lepore and Falk were able to examine—unlike all the researchers who had received tiny pieces of the brain from Harvey—was the outside appearance of Einstein's brain in its entirety. And then they were able to compare that with the average human brain. What they discovered was that the cortex of Einstein's brain was different from the average person's. One thing they found, for example, was an enlarged "knob" on the part of the brain that controls the left hand and is closely associated with people who play string instruments. (Einstein was skillfully adept at playing the violin.) "Every lobe of his brain on the surface was anatomically different from the human norm," Lepore told me, as he leaned into the webcam, half of his face taking over the screen. "This brain is not built the way John Q. Public's brain is built."

The two scientists published a first-of-its-kind paper in 2012, in which they explained that the surface of Einstein's brain had more folds and ridges than the average person's and that these additional elements may have been key to his creativity. The question of which came first—the proverbial chicken or the egg—remained: Was Einstein born with this special brain, or did coming up with theories of fundamental physics turn his brain into this unique specimen?

But even Lepore himself, who so desperately wanted to understand the basis for Einstein's genius and wrote a book about his research, questioned whether or not the surface anatomy of a brain is meaningful information. "You're telling me geniuses have different outside brains? Probably not so much," he admitted. "It's the internal connections." He instead pointed me to a study by Dr. Weiwei Men, a Chinese physics professor, conducted in 2013. Men discovered that Einstein had a colossal corpus callosum, the part of the brain wiring that carries information between the right hemisphere and the left. It's been described as a veritable superhighway of connectivity. Einstein's neural fiber network was so dense that it was thicker in key crossings than those of young and healthy men in their prime. Of particular note in the study is that the researchers compared the Einstein specimen to fifty-two brains of men aged twenty-four to thirty; Einstein was twenty-six years old in 1905, his "Miracle Year" when he published four groundbreaking papers that upended physics. Einstein's senior-citizen brain contained more connections than those of younger people. Another study found that Einstein's brain lacked the degenerative changes that would normally be present in a seventy-six-year-old.

Back in my kitchen, Carolyn tells me that she takes all these studies with a grain of salt. "None of this really answers the question: Was Einstein born with that brain or did experience shape it? If you spent the first years of your life reviewing patents for time pieces, maybe you start to think differently," she said, referring to Einstein's first job as a patent clerk in Switzerland. "You can keep growing new brain cells all your life."

She pointed to a 1984 study from Marian Diamond, a professor in California who received from Harvey four pieces from different sections of the brain. "Einstein appeared to have an average of 73 percent more glial cells than the control subjects," Carolyn said. But Diamond's study was widely criticized because she only compared it to eleven other brains, and she didn't know much about the people whose brains she used for comparison, including their IQ score and other information that would've been useful. It also was not a blind study. She knew which specimen belonged to Einstein, which could've colored her results. She also admitted to leaving out research that would've disproved her findings. Of Diamond's brain study, Carolyn wrote: "No one, including she, would have known a Mozart from a Manson."

So why are so many people, to this day, still enamored with Einstein's brain? "One thing that always occurs to me," Carolyn said, "is that the physical basis of intelligence or what makes one person smarter than the other has been an age-old fascination." She said it's also unlike other forms of bias. "Every other wall of discrimination has crumbled. You can get in trouble for not hiring somebody because they're gay or because they're a woman or because they're Jewish or because they're Black. But it is still acceptable and, in fact, expected that you don't get a job because somebody deems you not smart enough. So, it is still the last frontier of acceptable discrimination." She also thinks that, in the age of artificial intelligence, understanding the neural basis of intellect has taken on new meaning. "Because now we think that if we can figure out how the brain does it, then we can replicate it with computer networks. So, I think that fascination is not likely to decline."

One of the only places in the world where the public can see Einstein's brain is in downtown Philadelphia at the Mütter Museum, a neolithic building sandwiched between an office furniture store and a Trader Joe's. It was opened in 1858 by Dr. Thomas Dent Mütter with the hopes of expanding biomedical research and as a resource for the students at the College of Physicians of Philadelphia. Marie Curie, a close confidante of Einstein, visited the museum in 1921 and donated a piezo-electric apparatus, one of the first tools to ever measure radioactivity.

For most of the twentieth century, not many people knew about the museum. That all changed during the 1980s when Gretchen Worden, the curator, became a frequent guest on the *Late Show with David Letterman* where she showed off various creepy items from the collection—like the world's largest hairball or Victorian-era surgical tools. The museum went from just a few hundred visitors a year to accommodating sixty thousand tourists annually. After all, who wouldn't want to see a piece of tissue from the neck of John Wilkes Booth?

Its collection of twenty-five thousand medical artifacts is, to put it mildly, utterly bizarre. It's no wonder its slogan is "Disturbingly informative." As I discovered when I went there for a visit, the museum is a mecca of the macabre. More than one hundred skulls line one wall. Wax molds show how cancer spreads across a breast. An exhibit about the history of skin features life-size models of faces with gangrene, a nose littered with leprosy, and a baby's arm infected with the herpes virus. Look over there: it's a cancerous tumor from the mouth of President Grover Cleveland. I learned about Rapunzel syndrome, named after the Brothers Grimm fairy tale, a rare gastrointestinal condition resulting from compulsive hair eating. A large hairball forms in the stomach with the "tail" extending into the intestines. The largest one ever removed from a human weighed ten pounds and was surgically removed from an eighteen-year-old Chicago woman in 2007. (I may have thrown up just a little as I typed that sentence.) I can't wait to see what they sell in the museum's gift shop. Okay, I'll tell you now: They sell a postcard featuring a diseased colon that had forty pounds of fecal matter stuck in it. They also sold keychains.

I asked a docent where I could find Einstein's brain. I assumed the area devoted to the Nobel Prize winner would have its own room, or at least an entire wall, but instead she pointed me to a small corner. The last remaining cells of *Time* magazine's Person of the Century reside in an unassuming bookcase in the back of a room next to the glass coffin of the "Soap Lady" mummy, a body exhumed from a nearby cemetery in 1875. And, I should point out, Einstein's actual brain barely takes up 20 percent of the bookcase. And you can't even tell that it's a brain. What you're looking at is a box of forty-six microscope slides. On the slides are thin slices of the brain, looking more like a tiny smushed frog between two plastic covers.

Thomas Harvey had sent this particular set of slides to a lab at Philadelphia General Hospital where it eventually ended up in the hands of Dr. Lucy Rorke-Adams, a neuropathologist who studied brains. She remained, as the press described her, a "quiet custodian" of the collection for decades until 2011 when, at eighty-two, she donated them to the Mütter.

Einstein's is not the only famous brain at the museum. Just down the hall is what's left of the cerebral matter of Charles J. Guiteau, the assassin who shot President Garfield in 1881. Guiteau's long murder trial was one of the earliest to invoke the insanity defense. That legal strategy was not a success; he was found guilty and was hanged in 1882. However, doctors performing the autopsy saved the brain to conduct research, hoping to find some evidence of mental illness. They did discover some proof: small blood vessels, a trait shared by those living in asylums.

I walked through the museum's main lobby where a marble staircase led to a second-floor library as well as two grand ballrooms used for lectures and events. I can imagine a bar mitzvah in one while, in the other, a wedding couple snaps photos next to a vulva and dried penis. It was the end of the day, and the museum was closing. I snuck into the restroom where a sign above the sink promoted handwashing. It featured a drawing of a skeleton covered in soap bubbles.

Shortly before his death, Thomas Harvey decided it was finally time to find a new caretaker for his prized mason jars. Einstein's brain, shriveled and pickled, would live on. Harvey, alas, would not. So, he went back to the place where this story started: Princeton Hospital. Harvey hobbled up the stairs to his old office and found the person who now had his old job, the chief pathologist. He left the unassuming box from the Kaweah Citrus Association on the floor. It was a proverbial passing of the brain baton, from doctor to doctor. Harvey had finally unburdened himself.

I asked Carolyn the current whereabouts of the legendary jars. She told me the doctor Harvey gave them to still had them, tucked away for safekeeping. But she warned me that he was nervous to be in the public eye. He was afraid that crazy people— Einstein fanatics or relic hunters or Benyamin Cohen—might break into his home one night and steal the brain yet again.

She scribbled his email address on a piece of paper.

"Who knows?" she said. "Maybe he'll talk to you."

The next day, I shot him an email and, as expected, he brushed me off. But I was persistent and kept reaching out to him every few months as I was writing this book to check in to see if, just maybe, he'd be open to showing me the brain. Finally, after much back-and-forth and just as I was about to give up, he surprised me.

"How about Friday?"

"Wait, what?" I asked.

"Come over on Friday," he replied.

I thought perhaps he was joking. He lives in Princeton, a six-hour drive from where I live. I asked him for a hotel recommendation, we set a time for the meetup, and he hung up the phone.

And so that's how I found myself one cold winter Friday in

February, my head hunched into the trunk of a car, staring at a cookie jar filled with Albert Einstein's brain.

"This is Einstein's brain," he told me in such a casual way, it was as if he could be showing me a photo of his golden retriever. He held up a jar and pointed at what I can only describe as tofu bobbing around in brine.

"That's a piece of the brain, and that's another piece of the brain," he said. "There's the brainstem."

I got emotional. Twenty years earlier, the story of Thomas Harvey and the brain is what first got me interested in Einstein. (Spoiler alert: It wasn't the quantum mechanics.) And for all these years, it had just been this story written in a book. And now, here I was holding Harvey's actual jars. But more than that, I was holding Einstein himself. Sure, he was dead, his body cremated. But somehow, through some twist of fate, I was able to reach across time itself and hold a piece of history. This was an actual piece of Albert Einstein's body. I started to tear up. I thought about all the new ideas this brain imagined. I had embarked on a journey to find all that this genius had inspired and here I was holding the original relic itself. This was more than just the physical remnants of a famous person; Einstein's brain represented so much more: the curiosity that surrounds our very existence in the universe. There's a mystical aspect to the brain. When Dr. Witelson examined it, she described a feeling of awe just to be near it. Steven Levy, the journalist who first brought the story of the brain to the masses, called seeing it a "religious experience."

The doctor who keeps the brain of the world's favorite genius in his car does not want to be named. That was his one condition for this meeting. I could only refer to him as Dr. X in this book. After all, he didn't want the whereabouts of the brain to become public. "Harvey gave it to me. And I've kept a very low profile, at least I've tried," he said. "And I've never made a dollar off of it because he wanted it to be used for scientific purposes. I'm very reticent to be out there with it."

Dr. X cuts a tall and trim figure and is fit for a man in his late sixties. He's dressed fashionably in dark tailored jeans and a button-down shirt covered in a blue pullover sweater. He's Larry David, but with more hair. And in case you're thinking about tracking down the brain yourself, you should know that it's not always in his car. Otherwise, thieves could just troll around Princeton looking for a dark blue Acura MDX with a weird odor wafting out the back. It's mostly out of sight, out of mind—kept in a filing cabinet in his office or under the floorboard at his home, he wouldn't tell me. About once a year, he opens the jars to replenish the alcohol inside for preservation purposes.

As the new caretaker of the brain, he doesn't give much credence to the scientists who have come before him. When I bring up Witelson's study that showed how Einstein's brain was unique, he scoffed. "That was a bullshit paper," he told me matter-of-factly. "She knew which brain was Einstein's. She got a lot of notoriety about it." He paused. "She got a lot of shit about it, too."

Dr. X, in many ways, has become Dr. Harvey 2.0. Like his predecessor, he's had several offers for the brain, but he is dubious of those prospects. He told me he's had interest from a brain bank at Harvard and has taken calls from the Smithsonian. But he's afraid what will happen to it there. "All I ever think about is *Raiders of the Lost Ark*," he said, reminding me of the movie's final scene in which government officials pack up the ark into a crate and store it in a vast warehouse full of other crates. "You'll never see it again," Dr. X. said.

"If and when I ever give it up, I have no idea what I'll do with it." He said he might leave Einstein's brain to his kids, but they don't seem that interested. I want to shout, "I'll gladly take it," but I couldn't muster the words, or the gall.

Dr. X has shown pieces to his brother-in-law, a molecular biologist, to see if Einstein's DNA could be studied. But it was determined that the brain was not preserved properly in 1955, and that the DNA was too fragmented to do much with. In other words, it would be nearly impossible to clone Albert Einstein. But Dr. X holds out hope. "I guess at some point in the future, maybe it will be like *Jurassic Park*," he said, referring to the fictional story of bringing back dinosaurs from extinction using genetic material. "Maybe we could put the pieces back together." Cloning Albert Einstein had never crossed my mind, but now it's all I can think about.

A team of Harvard scientists had made headlines in recent years for using genetic engineering to help revive the extinct woolly mammoth. If the species returns, the animals could live in Pleistocene Park, a wildlife preserve being built by a father–son duo in the frozen tundra of Siberia. So what's one brain compared to an entire species? Could Dr. X have the hubris to actually pull it off? Even if not, as Einstein's brain eventually gets passed down from one generation to the next, it has, in a sense, itself become immortal.

Dr X. slammed the trunk closed, and we hopped into the front seat. He revved up the engine; the satellite radio instantly came to life with the latest news from MSNBC. It's at this point that I should probably tell you where we were: we were in the parking lot of the Princeton Airport, an airfield so small there is just one runway. It's home to a flying school, and it's where Dr. X has a hangar that houses his tiny plane.

So why did I throw up on the day I saw Einstein's brain? That story takes place up in the air three thousand feet above the ground—and involves the search to find how a theory Einstein came up with a century ago helps us each and every day.

FLYING, DRIVING, AND SURFING WITH EINSTEIN

"Thanks to my fortunate idea of introducing the
relativity principle into physics, you now enormously
overrate my scientific abilities, to the point where
this makes me quite uncomfortable."

—ALBERT EINSTEIN

I n case I die, here's how you release the parachute."

Those are not the words you want to hear from anyone, let alone the pilot you've just entrusted with your life. But at that point, it was too late to turn around. Dr. X and I were cramped into a small four-seater propeller plane with the engine revving at the start of a runway that would test my belief in the power of gravity. Dr. X was manning the controls, flipping switches, turning thingamajigs, and going over a safety checklist. One that, apparently, included scaring the bejeezus out of his passenger. I glanced around the cockpit, feeling queasy at the notion of my pilot's potentially imminent demise. The Cirrus SR22 GTS has a nickname, "the plane with the parachute," because instead of your seat ejecting and floating gently back down to Earth, a parachute blasts out above the *entire* plane. At that point you just have to hope this glorified tin can, hurtling at a speed of about two hundred miles per hour, somehow decides to slow down before reacquainting its fuselage with terra firma.

"When it lands," he added, "run out quick."

After all, he helpfully explained, planes tend to explode on impact. Despite all of this, and perhaps naively, I felt slightly reassured simply because Dr. X looked straight out of central casting: a bomber jacket, a fancy gold watch, aviator sunglasses, and a mane of silver-fox hair. He attempted better bedside manner by informing me that my first-ever flight in a small plane would be on the luxurious "BMW of single engine air-craft." Indeed, it had tan fur-covered seats and satellite radio. But still, listening to commercial-free Bob Dylan wasn't going to make an impromptu landing any softer. On the upside, it was beautiful outside, with clear blue skies as far as the eye could see. A good day to die.

The engine was so loud at this point that we were commu-nicating via over-the-ear headsets even though we were sitting right next to each other. He guided me through the "in case of emergency" sequence.

"Don't pull the parachute handle until you do this," he said, all-too-quickly running through a series of knob turns and key clicks and lever pulls and something called a straight and level button—"For God's sake," he warned, "whatever you do, don't forget the straight and level button"—all of which were some-how supposed to save me in case he had a heart attack while flying over the Delaware River (the same place where Einstein's family reportedly disposed of the genius's cremated ashes). He pointed to my feet and showed me the backup brake I would need to slam. There was, however, one glaring problem. I was too short to reach the pedals.

"Um," he said looking at me and then looking down at my dangling feet. "It's okay. Hopefully, you won't need to do any of this."

The reason for this entire exercise—me putting my life in the hands of a pathologist who happened to also have a pilot's license—was, of course, because of Albert Einstein. I mean, at this point, I'm pretty sure you've figured that out. That's because Einstein is responsible for one of the most ubiquitous modern marvels known to mankind: the Global Positioning System, or as it's better known to you and me, GPS.

Unbeknownst to most people using Tinder or Instacart or Waze, the theory of relativity serves as the backbone for GPS. Einstein explained that even though a satellite in the sky can be moving through space at the same time as the Earth itself is rotating, we can still have a consistent frame of reference because of the concept of relative motion. The next time you have a pizza delivered to your home, thank Einstein. GPS is everywhere—from cars to cell phones. It helps us navigate through traffic and get to work each day. It powers our fitness watches that keep count of our daily steps. It makes it possible for us to track our Amazon Prime packages. It allows us to swipe right and find a nearby date (or a hookup) for the evening. I even got our Great Pyrenees, Jolene, a GPS dog collar. In case she were to ever run away, we'd be able to track her down. And it's all because of calculations conceived by Albert Einstein.

I thought who better than Dr. X, uniquely positioned as both an amateur pilot and the current caretaker of Einstein's brain, to show me how he uses GPS in the sky. So, I buckled up. It was time for takeoff.

Google the inventor of GPS and two names will pop up: Albert Einstein and Gladys Mae West. I assume the latter will be unknown to you, so allow me to tell you her story: West was born in 1930 to a family of sharecroppers in rural Virginia. Her family was Black and poor; her dad was a farmer, and her mom worked in a tobacco factory. Growing up, West had one guiding principle: she didn't want to work on a farm. She quickly realized that the only way out was an education. She was a good student, but college tuition was too far out of reach. Fortunately for her, she was her high school's class valedictorian and, in 1948, she received a scholarship to attend Virginia State College. She used babysitting money to pay for room and board. She earned a math degree and spent a few years teaching after that. But then her life would forever change when she got a job with the U.S. Navy.

This was back in the days when people were hired to help engineers by becoming human calculators. "I did not start out using the big computer," she told me when I reached her in the spring of 2022 when she was ninety-two years old. Other Black women of her generation, who worked similar jobs at NASA during the space race, were later immortalized in *Hidden Figures*, a book and movie of the same name. West's job was to analyze the data coming from satellites, specifically seeking to determine their exact location as they transmitted signals across the globe. She worked her way up the ranks onto a team of five people working on a special project. In the early 1960s, as Einstein's brain bobbed around in a jar in Thomas Harvey's office, West and her colleagues authored a landmark study that proved "the regularity of Pluto's motion relative to Neptune." In other words, they had unknowingly invented the modern-day GPS. "The world was getting bigger and different," she said, "and I was a part of it."

Think of GPS as a form of triangulation. Let's say your car is moving at sixty miles per hour and headed west. Your phone sends those coordinates to the satellite, which in turn, sends back information on where you need to go to get to your desired destination. That would be fine and dandy except, as Einstein pointed out in his general theory of relativity, clocks on a satellite in space run slightly faster than clocks back on Earth. In essence, due to gravity and the Earth's atmosphere, there is a slight time delay from when the satellite sends the signal to when your GPS receives it. So, Einstein created a simple and elegant mathematical model to make up for that difference to keep us on track. Even a small miscalculation could mean the difference between staying safe and your car drifting over into a lane of oncoming traffic. Our devices calculate the time difference between receiving signals from different satellites to pinpoint our location on the Earth's surface to an impressive degree of accuracy—within a few feet of where we're standing at any given moment. "Without Einstein's theory," writes Robbert Dijkgraaf, a Dutch physicist, "the GPS in our smartphones would drift off course by about seven miles a day." It would be a massive error that would render GPS useless.

West's early research used complex algorithms—accounting for gravitational pull and the shape of the Earth—to translate Einstein's theory into reality. "I remember the time when his theory was discovered," West told me. "Right off, I didn't quite understand it, because I didn't have an application to put it to use. But as I did more projects, the theory became more applicable to what we were doing." When you turn on your phone's GPS and walk a little in one direction, Einstein's theory plus West's calculations figured out that you're now exactly so many feet away from where you began. "The world was excited about it," she said. "I was excited to be living in a time when he made the discoveries. I felt very proud of him. All I did was honor and support him."

West's work served as the basis for GPS, which was originally developed in 1973 by the U.S. Department of Defense with twenty-four satellites. Until the 1980s, regular civilians couldn't even use the service. Even to this day, GPS is still managed by the U.S. government, and it can decide to selectively deny access to anyone, as it did with the Indian military during the Kargil War in 1999. (India has since designed its own system, called the Indian Regional Navigation Satellite System, or IRNSS for short. Other countries, like Japan and Russia, have built systems that work as add-ons to the U.S. GPS system.) Now GPS has become second nature for all of us, as we drive around, following a blue dot slowly across a map showing us exactly where to go. But West, despite her invention, still prefers to do her own calculations. You see, Gladys West doesn't use GPS. She's a fan of old-fashioned paper maps. "I'm a doer, hands-on kind of person," she admitted. "If I can see the road and see where it turns and see where it went, I am more sure."

Three thousand feet above the Princeton Airport, Dr. X took his hand off the steering wheel. "We're on autopilot now. I'm not flying," he said, as if trying to make me feel more at ease. A vast brown, winter landscape lay before us—highways and homes and a water treatment facility. The plane had a built-in Garmin GPS monitor on the dashboard, similar to what you'd see in a car, but Dr. X preferred the interface of the ForeFlight app on his tablet. It's basically Waze on an iPad. He punched in where we were and where we were going—an hour jaunt toward New York City and back around over small Pennsylvania towns, before landing back in Princeton.

He told me that pilots stay on course by using waypoints throughout the sky, which are based on GPS coordinates. Ten minutes into our flight we reached the Solberg waypoint, a marker only visible to pilots. Next up was Lanna. Each one corresponded to a different latitude and longitude of a particular location.

"Midair accidents never happen because of GPS," he told me, and then added: "Well, rarely." Prior to the GPS, pilots used something called the Instrument Landing System (ILS), which offered ground-based guidance to help planes land. Nowadays, GPS is so advanced that it offers pilots precision on lining up with the runway. GPS even adjusts for wind.

All this talk of wind wasn't sitting right. Literally. Somewhere over Allentown, Pennsylvania, I began to feel queasy. I had stupidly eaten a breakfast for three Benyamins that morning, and regret was washing over me. I kept looking off in the distance, trying to tell myself we'd be down on the ground soon. But as he calmly flew us around, zagging over the religiously named Nazareth and Quakertown, I realized we hadn't even arrived at the halfway point. I quickly asked Dr. X if he had a barf bag on board. When he realized I wasn't joking, he pointed me to a small sandwich sized Ziplock bag filled with face masks. (We were still in the midst of the coronavirus pandemic.) I haphazardly dumped everything in the bag by my dangling feet and proceeded to throw up two eggs, some banana pecan pancakes, and a cup of coffee, heavy on the cream.

"Sorry," he said, looking at me with both empathy and derision. (I had, in essence, just puked in his BMW.) "This is one of the bumpier trips I've ever had."

Luckily, my pilot was also a doctor. I'm sure he'd seen worse. To pass time, and I guess to make me feel better, he explained what my body was going through. It had nothing to do with the food I ate that morning, he said. It was the vestibular tubes in my ear sending mixed messages to my brain: my body was both moving and sitting still at the same time. That kind of confusion causes nausea and vomiting. He told me that he threw up the first time he did an autopsy. They had to put him on a gurney and wheel him out of the room. "But just that first time," he offered with a sense of pride and duty. "After that, I had a job to do, and I had to concentrate."

I closely watched the GPS as we started making our way

back to the airport. As we got closer, we made a pass over Einstein's house at 112 Mercer Street, about a mile from Princeton University. Earlier in the day, I had walked to the home, which is now a private residence with signs urging people not to trespass. I walked back and forth on the sidewalk in front of it, pretending to be on a phone call and—when I thought nobody was looking—I took several selfies. After Einstein's death, his stepdaughter Margot lived there until her passing in 1986. The home is now owned by the Institute for Advanced Study where Einstein worked. It's rented out to visiting professors: several Nobel Prize winners have lived there over the years.

We landed and I've never been so happy to step foot into New Jersey. I quickly disembarked and sat down on the tarmac, hyperventilating while Dr. X parked his plane back in the hangar.

After my midair adventure, I figured I should try a land-based activity to learn how Einstein's GPS is helping us all navigate our lives. So, I googled "How to become an Uber driver." In my day-to-day life, I don't often need to use GPS when I'm driving. I usually remember how to get from my house to the grocery store and back home again. If I wanted to really see the benefits of GPS in action, the best way would be to force myself to drive to new locations. That's why I decided to become an Uber driver.

GPS first became available in cars in the 1990s, but you had to be a Saudi prince to afford it. And even those who spoiled themselves had trouble navigating: When BMW first introduced GPS in its luxury Five Series car, German men refused to take directions from a female voice, wrote Professor Clifford Nass, an expert on human-computer interaction. For most of us, the ubiquity of in-car navigation took hold in the first decade of the 2000s and really accelerated with the 2007 introduction of the iPhone. Although, it should be noted that the built-in Apple Maps program was notoriously unhelpful when it debuted. There were countless news reports of people ending up in the middle of nowhere after following its directions. Entire cities were sometimes missing from the map, parks were often listed as airports, and the instructions might tell you to keep driving directly into a body of water. Those kinks were sorted out, and eventually we end up in our present day with NASA-grade technology now in the palm of our hands.

Uber takes the concept of GPS to a whole new level. Instead of using it to get directions for yourself, you're using your location to have someone else find you. The service is an ingenious idea: Someone needing a ride pushes a button on the app and GPS instantly pinpoints their position, alerting drivers in the area. Drivers receive a notification on their phone, and with a simple swipe, can decide to pick up the passenger. The passenger then types in where they want to go, and the driver uses Uber's built-in GPS to drive the passenger there. Payment is made using a credit card on file. It's a completely frictionless process (the two people never even need to speak, aside from the initial pickup when the driver confirms the passenger's name). Uber launched in 2011, first in San Francisco, and eventually across the globe. By 2021, it operated in more than 10,500 cities in seventy-two countries. And it's all made possible because of a mathematical solution devised by Albert Einstein.

As I soon discovered, it turns out the only requirements to becoming an Uber driver are having a license and a car, which is both satisfying and scary at the same time. Think about that the next time you hop into the back of a stranger's Honda Civic. Uber's website claims you can sign up in minutes, so I started the process. When I opened the app for the first time, it showed me photos of other drivers—a collection of young, old, diverse, always smiling people who apparently had both a car and a driver's license—a community that beckoned me to join its ranks.

I watched the dozen different welcome videos. One video told me to jazz up my profile with quirky details, so I added that I love to watch reality TV. Who knows? Perhaps I will pick up a fellow fan of *90 Day Fiancé*.

I got my car washed, filled up the air in the tires, and put a couple of water bottles in the holders facing the backseat of my pickup truck. I turned on the app and waited. As my wife Elizabeth helpfully pointed out, we live in the middle of nowhere; the likelihood of me getting a pickup request was pretty slim. There are more cows than cars on our road. Indeed, the first thing I noticed when I got in the car and opened the app was that Uber couldn't even locate where I was. My house, ironically, is not findable on GPS. So, I drove into town, with the app running in the background, sending out a beacon to the tired, the poor, the huddled masses that I was available to drive them wherever they wanted—within reason.

I wasn't the first person, nor would I be the last, to moonlight as an Uber driver. There are approximately one million rideshare drivers in the United States. One of them is former Senator Ben Sasse. The Nebraska Republican occasionally hopped behind the wheel as a way to stay in touch with his constituents. Another is Elwood Edwards, the voice behind AOL's famous "You've Got Mail" greeting. With not much else to do since everybody stopped using AOL in the 1990s, Edwards is now an Uber driver. (As videos on social media can attest, he loves telling his passengers about his previous career.) Lyft, Uber's main competitor, had a marketing campaign where they filmed celebrities in wigs going undercover as drivers. Why yes, that *is* Shaquille O'Neal taking you to your dentist appointment. Einstein, who never bothered to get a driver's license, would have likely appreciated a ride from one of these chauffeurs.

Uber, like so many other aspects of our daily existence that we take for granted, seems unimaginable to a younger version of myself, let alone someone living in the era of Einstein. Growing up, when my family would take road trips, we called up AAA, told them where we wanted to go, and they would prepare a "TripTik" map just for us. And then they would mail it to us. And we would wait. The modern-day GPS means we can decide to drive to Albuquerque, New Mexico, at a moment's notice.

I live in a college town, which is both a blessing and a curse for an Uber driver. On the one hand, there is a huge pool of driverless people who need rides. On the other, there's a pretty good chance you'll be chauffeuring around someone with a hangover or, worse, who throws up on your seats. For this reason, I chose to limit my driving time to early afternoon hours. I figured that, by then, the hangovers had ended, and the nightly drinking had yet to begin. I've lived here for a decade and know my way around. Let's be clear: the town is only ten square miles. Yet, when the app blinked awake with my first request, my heart skipped a beat. I was both nervous and excited.

I pulled up to a classroom building on campus and a student chatting on his cell phone hopped into my backseat. He hung up and we started talking. I told him he had the distinction of being my very first passenger, which didn't seem to impress him. I dropped him off at his dorm. The entire trip, according to the app, took ten minutes. I earned $5.97. I was already at the dorms, and a minute didn't go by when I got another request. A student needed a ride to her car, which was parked in a lot two miles away. It took seven minutes, and I made $5.74. The next trip was the silliest. It was a student who basically wanted a ride across a busy street. It took four minutes (I made a wrong turn) and I made $4.17. She threw in a $3 tip for my trouble. What I started to realize was that people here don't need to go far; they're just lazy.

My next request for an Uber came from the hospital. I wasn't too far away, but I missed a turn, and it took me an extra few minutes to get there. I guess Einstein didn't consider roundabouts when conceiving GPS. A nurse brought down a patient in a wheelchair, a woman in her forties, to the front entrance, and we helped her climb into the truck along with a couple of suitcases. She had spent a week receiving treatment for a rare medical disorder, and she was headed back home. She lived in a small town in another state, and I was dropping her off at the Morgantown Municipal Airport.

My passenger, Michelle LaBar, has complex regional pain syndrome. "Your neurological system goes haywire, and becomes super sensitive," she told me. "So it interprets everything incorrectly. Like a little bump, it interprets as a major injury." I took a mental note and tried my best to not drive over any potholes.

One of the only hospitals in the country that offered an experimental treatment was in town at West Virginia University. She lived a six-hour drive away. That's when she heard about Patient Airlift Services (PALS), a non-profit organization of volunteer pilots who arrange free flights for people needing medical treatments.

LaBar told me that she has become reliant on GPS. "Being disabled and with the ketamine infusions, I have a lot of memory problems. I couldn't make it to familiar places. I've gotten lost going to the doctor's office." Now, remarkably, she wants to give back. "There's a lot of people in my situation. We rely on it so much." She signed up to be a driver for Lyft, Uber's main competitor, but all the paperwork was too much for her. She still plans on helping anyway. "One of my goals is to volunteer a little bit," she said. "I just want to be able to give little old ladies rides to the grocery store." And so it goes full circle.

We arrived at the airport, and I helped carry her luggage up a flight of stairs and out to the tarmac. A technician was gassing up a small Cessna and the pilot came to greet us. It was December, and he was dressed as Santa Claus. Here we all were, three complete strangers, who had each used GPS and ended up working together to bring a sick woman back home.

We exist at this unique crossroads in time when adults can remember life before and life after the advent of the internet. Eventually it will become like television, or the radio before that, or the printing press before that—nobody will be around to remember what living without it was like. "If we're the last people in history to know life before the internet," Michael Harris wrote in *The End of Absence*, "we are also the only ones who will ever speak, as it were, both languages." He calls us a part of the "straddle generation, with one foot in the digital pond, and the other on the shore."

There is certainly something we've given up by outsourcing the bulk of our knowledge to GPS devices and Google. Remembering someone's phone number or writing down a detailed list of directions now seems as foreign to us as commuting to work by horse and buggy. There is simply no need. "Should we ever bother to memorize poetry, or names, or historical facts?" Harris asked. "Fifty years from now, if you have an expansive, old-fashioned memory—if you can recite *The Epic of Gilgamesh*, say—are you a wizard or a dinosaur?" The quote, "Never memorize something that you can look up" is often attributed to Einstein. But I did look it up, thanks to the internet, and Einstein never actually said that.

Shortly after the start of the COVID-19 pandemic, when work from home was becoming the new normal, the internet stopped working. I mean, not everyone's internet, but *our* internet ceased operating. That beloved green light on the modem turned a scary red. While everyone else was Zooming in to important meetings, wearing work clothes above the waist and pajamas down below, the internet at our home just conked out.

Where we live, deep in the woods in rural West Virginia, there is only one company that offers internet service. And by "offer internet service" I'm being generous. They only provide DSL speed, which was all the rage when Bill Clinton was president. Nowadays, not so much. When I had large files that I needed to send online, I would start the process before I went to bed and hope that it had all uploaded by the time I woke up in the morning. It was far from great, and we had gotten used to it. But now, we didn't even have that. Apparently, a storm knocked down a tree and that tree knocked out the cable to my entire neighborhood.

If you're thinking, *Surely, they'll come out to fix it posthaste,* I wish I could bottle up that naive optimism of yours. The problem is when I say the storm knocked out the internet for our entire neighborhood, I should clarify that in such a rural area, that consists of only twenty people. Needless to say, our outage was not really of top concern for our internet provider. I won't mention their name, but it rhymes with Shmontier Communications. Screw that, it's Frontier Communications. Let's just say, I completely understand why cable providers are some of the most loathed companies in the country. Comcast landed on a recent customer satisfaction survey of the most-hated corporations in America. Also on that same list? The Weinstein Company and the Trump Organization. You do the math.

Granted, it's not all Frontier's fault. For the first two weeks, the company didn't realize a tree had fallen. *If a tree falls in the forest, and your cable company isn't there to hear, did it actually fall?* When our entire neighborhood called customer service to complain—and were rerouted to a call center in India—the company's response was that the problem must be with us. Yes, they really thought that by some bizarre hi-tech twist of fate, every modem on our street had mysteriously died on the exact same day. They told us to reboot them, to reset them, to unplug them, and do a little dance. When that didn't work, they sent out new modems. More days went by, and still no internet. Keep in mind, I was working from home even before the pandemic began; my livelihood depended on the internet. After all, how could I post status updates and memes to Einstein's social media pages if I didn't have access to the web? So, I grabbed my laptop and drove to a McDonald's parking lot and tapped into the restaurant's free Wi-Fi. I sat in my car, too afraid to go inside for fear of being exposed to a deadly virus. Yes, when you saw that hilarious tweet from Albert Einstein, it was me dialing in from the back of my pickup truck with a golden arch in the background.

We tried explaining all of this to the geniuses at Frontier, but each day we'd end up speaking to a new case manager who said not to worry, they were working on it. That runaround lasted more days. I had finally had enough. I emailed local reporters and told them what was going on. We live in a college town where practically everything revolves around the university, where my wife is a professor. In an effort to keep everyone safe during the early days of the pandemic, the school had recently moved all classes online. The university touted how smooth the transition had been. And yet, Frontier was preventing one of the professors from teaching. That framing of the issue (as opposed to the guy who tweets as Einstein having to drive to a fast-food parking lot) caught the attention of the media. A local TV news crew showed up at our house the next day. Our story made the six o'clock news, with the chyron dramatically billing it as "Fighting for Connectivity."

The least surprising thing happened the next day: We got a call from Frontier's headquarters saying they saw the news program and would start taking our issue seriously. They finally cleared the tree debris and got our internet up and running again.

Going without internet for more than a month will change a person. Sure, going to prison will also change a person. But even prisoners get email. Living a modern life cut off from the rest of the world is not living. I decided I would never let this happen again, so I set forth on finding an alternative internet provider. The problem, as I said earlier, is that living where we do, there aren't any options. It's not like I could call Time Warner, or Verizon, or heck, even the much-reviled Comcast. What I wouldn't give to have Comcast service. But alas, the closest Comcast tower is two miles from my house, and I'm almost certain they don't make extension cords that long. Although, as my wife pointed out, we probably could daisy-chain a bunch of cords together. But we wouldn't have to stoop to that level because this is when Elon Musk, one of the world's richest people, came to my rescue.

Elon Musk has been called an Einstein of our times in that he thinks outside the box and has come up with new concepts that have changed the world as we know it. Musk has infiltrated challenging industries—from electric cars to rockets to solar power—and leapfrogged other companies with transformative advancements. Case in point is his Starlink company. High-speed internet works through a central facility, and wires from that building are spread out across the city, like tentacles from a spider's body. The only major advancement since the 1990s, at the dawn of the consumer internet, is that providers are moving away from attaching these lines to telephone poles and instead burying them underground. But the basic concept still remains: you have access to the web because your home modem is tethered to a wire.

But what about people like me, who live in the forest, where dragging a wire from home to home can be cost-prohibitive? Or what about people who live in third-world countries where no infrastructure exists for wires and poles and fast internet? For the answer to these questions, Musk looked to the sky above. Like Einstein, he questioned established doctrine: Why do we even need poles and wires? That's an outdated mode of thinking. Instead, Musk proposed an entirely new paradigm: Couldn't we get the internet beamed down to us from satellites in space? It's not so far-fetched. People with DirecTV or Dish Network dishes attached to the roof of their home already get TV from space.

Fortunately for Musk, he had the infrastructure in place to test his theory. His SpaceX company was already sending rockets to space on an almost weekly basis. He could load one of them up with relatively inexpensive satellites and release them into the sky. Anyone who lives near one of those satellites can sign up for access to Starlink internet. And so that's exactly what I did, when I became one of the first beta testers for Musk in the summer of 2020.

The secret sauce that makes Musk's service work is, like GPS, rooted in Einstein's theory of relativity. Consider the following scenario for someone who has typical broadband internet in their home: You do a search for "Mother's Day gifts" on Google. As soon as you hit Enter, a signal dances down the chain of wires and telephone poles and eventually ends up at your internet provider who, in turn, quickly sends the Google results back to you. This all happens pretty much instantaneously since your modem is tethered to a wire. But when you're accessing the internet through space, there are all sorts of other calculations that come into play.

Most satellites are so far up in space you can't see them with the naked eye. Beaming signals back and forth takes time. You will experience what's called "latency" between the time you start your search for Mother's Day gifts and when the satellite's response ultimately arrives back at your house. Or imagine being on a Zoom call and you say hello, but the person on the other end doesn't hear your salutation for several seconds. Musk's solution was to put his Starlink satellites into low orbit, so low in fact you can see them from your driveway. Some backyard astronomers are already complaining that Musk's satellites are obstructing their view of the night sky.

Since the Starlink satellites are physically closer to Earth, there is less of a lag from when you send your request for information and when you get it back. But latency still exists, especially since Starlink satellites float around during the day. Not to mention that the Earth is constantly rotating as well. At 9:00 a.m., the Starlink satellite may be floating above the right side of your house, and at 4:00 p.m. it may be all the way over on the left side. The satellite's constant movement means that it is always at a slightly different distance from my house. What's more, Starlink has now introduced a service where you can have satellite internet on-the-go, like if you're driving in a Winnebago down the highway.

The only way for it to work seamlessly where you don't notice any latency is by using the theory of relativity that we mentioned earlier. Einstein's mathematical equations figure out the amount of time it takes for the signal to beam back and forth, as well as accounting for the effect of gravity over the atmosphere on that signal. It allows for Starlink to refine its measurements, so the end-user doesn't notice any lags.

This technological advancement—being able to provide high-speed internet to anyone anywhere—is nothing short of game-changing. Within days of Russia invading Ukraine in the spring of 2022, Musk sent up a few satellites above the war-torn country, so Ukrainians could still have access to the web even if all the local internet providers got shut down or were bombed. As of this writing, Starlink is now available in thirty-two countries across the globe. The company has launched more than 2,500 satellites into space, with plans to launch 40,000 more. That's five times the total number of satellites we've launched since the dawn of spaceflight. And the calculations that make it possible stem from GPS and Einstein.

GPS, it should be noted, is far from perfect. I once went to Berkeley Springs, West Virginia, to visit George Washington's bathtub. (It's a thing. Google it.) My GPS led me to what I thought was the parking lot, but instead my truck was staring at the side of a mountain. That's when I noticed a big yellow sign a bit too late: TURN AROUND. DEAD END. YOUR GPS IS WRONG. A friend in Baltimore lives in a row house that backs into a tight alleyway where you can't turn around. He told me that the locals know this and so never drive down there. But GPS continues to send delivery drivers to that route. My UPS delivery guy must be cut from a different cloth. When it snows, making our long driveway impenetrable, he simply texts me and I walk up to meet him, and then I sled back down to my house, package in hand.

The ubiquity of GPS, which shows us every back-alley way to get from Point A to Point B, often means we're avoiding the main roads. But those routes are there for a reason, devised by urban planners to go past retail stores and gas stations— and avoid driving through serene suburban neighborhoods. "Formerly quiet residential streets are filled with late-night Lyfts hauling drunken teenagers," writes Pamela Paul in *100 Things We've Lost to the Internet*. Furthermore, a team of scientists from the Czech Republic found that using GPS makes us have a worse sense of direction.

One of the obvious hiccups we all face when using GPS is that if we go into an underground tunnel or drive deep into a forest, that reliable blue dot we like to follow on our screens sometimes stops moving. GPS simply cannot work through obstacles like mountains or buildings. Without visible access to the sky, the satellites in space become weaker, and the signals they send can take valuable time in reaching your device. Dr. Lia Li, who graduated from University College London in 2016 with a PhD in quantum physics, has devised a clever solution— one that doesn't wholly rely on the signals from space.

Li was born in China in 1988 and raised in England. Growing up, she thought she wanted to be a lawyer. "It came from wanting to investigate stuff, seeking truth, revealing things that you didn't know before," she told me when I reached her on Zoom at her home in Bristol thanks to my now-working internet. She's got oversized black glasses and has dyed her hair pink. "I was always attracted to working in a job that allowed me to be curious."

She chose science over the law because she wanted to experiment with her hands. It's the same reason she, like Einstein, enjoyed playing both the piano and the violin. "I think that's when I really tuned into the connection between mind and hands, where you realize that you can physically do things that you never thought were possible." She focused in on physics because, as she framed it, "science isn't fixed. There's plenty more to be discovered. To this day, you can redefine many aspects of physics. It's not a stagnant field but one that's constantly evolving."

She said her North Star was Einstein, who worked in so many different fields—quantum mechanics, Brownian motion, the behavior of light, and on and on—that on the surface appeared to have very few, if any, connections. "It inspired me to step out beyond my comfort level," she told me. "I've always thought of him as a visionary, as someone who is not afraid to suggest ideas. He was just very curious and wanted to uncover more truth."

Four years after finishing her PhD, in 2020, Li stumbled upon such a truth. She had been fiddling around with optical inertial sensors, which measure acceleration and rate of rotation to keep track of a moving object without the need for external reference points. In other words, they provide the navigation aspect of GPS without the need of a satellite. Granted, you couldn't use these sensors for an entire trip from, say, New York to California, but they would certainly prove useful for those intermittent times when you're driving and your car's GPS receiver can't connect to the satellites in the sky—like in a tunnel or the forest.

I told Li about an app my wife uses when she goes on remote hikes. Elizabeth doesn't have access to GPS while in the woods, so to prevent herself from getting lost, she downloads a map of the area before she embarks. Li told me her technology offered a more refined solution. On top of the inertial sensors, she's added an algorithmic layer. It has the ability to instantly determine Elizabeth's walking speed and gait and refine how it tracks her as she's moving. This kind of tool would be useful in any kind of indoor setting where GPS is not available—like an airport or a shopping mall or a big stadium—where someone is trying to get from Concourse A to Concourse D. "I certainly get lost when I'm in an airport trying to make my connecting flight," Li said with a laugh.

Inertial sensors have been around for decades and have been helping people who have limited access to GPS travel safely, like on submarines and in rocket ships. But Li has figured out a way to bring that performance down to the size of a computer chip that can be built into a smartphone and available to everyone. She launched a startup, called Zero Point Motion, which aims to do just that.

Sure, having proper navigation while in a tunnel is certainly helpful, but Li sees a more revolutionary application: driverless cars. To keep riders safe, autonomous vehicles need every tool at their disposal—GPS, emergency brakes, 360-degree cameras, and whatever else engineers can stuff in there. Having Li's tech built in would extend the safety of riding in such a car. "It's not about necessarily replacing GPS," she told me, "but it's about complementing it." And it's not just in forests. Driverless cars have a tough time in dense areas like New York City, where GPS signals can bounce around between two tall skyscrapers, thereby extending the time it takes for the signal to get back to the car or degrading the signal strength. "It's what's known as the *urban canyon effect*," she said.

Li's technology is rooted in Einsteinian physics. "Without Einstein's contributions to GPS, notably his work on special and general relativity, we'd all be out of sync with each other," she said. Scientists are now at work on a GPS system that goes beyond Earth. It's called Galactic GPS. Imagine satellites hovering around the Moon or Mars offering driving tips to autonomous vehicles on those planets. "My perception of Einstein," Li said as we finished our call, "is this visionary leader who lived ages ago, who had all these ideas that are much bigger than even he realized. To think that his fundamental theories are now real technologies is amazing."

Back inside a hangar at Princeton Airport, Dr. X unfolded two reclining lawn chairs and tossed me a bottle of water. Not surprisingly, washing out the barf aftermath felt good. As did lying down and taking in deep breaths of fresh air. But my serenity didn't last long, as Dr. X began regaling me about the many ways a plane could crash.

"Running out of fuel is a dumb reason for crashing," he said to me, matter-of-factly.

"What would be a good reason to crash?" I asked.

He said one of the chief causes of accidents is pilots who don't have enough practice using the instruments on board and instead rely solely on their eyes. "They haven't had enough training to understand what to do in the middle of clouds," he said. "They can get disoriented. And before you know it, you're in a nosedive to the ground." According to the National Transportation Safety Board, this is what caused John F. Kennedy Jr. to crash his plane in 1999. "You've got to trust your instruments," Dr. X explained. "You don't trust your body."

Dr. X reminded me of the all-important straight and level button, the one that he made me promise I would push if he suddenly had a heart attack mid-flight. It engages an auto pilot function and he said he used it often. "It gives me a chance to get my thoughts together, my maps together, and not have to worry about flying the airplane." It is, like GPS, a technological guide to help when needed.

The irony, of course, is that Einstein—the guy who laid the groundwork to make it possible for computers to pinpoint our location—was himself notoriously bad with directions. Indeed,

Einstein never learned to drive. Even while sailing, one of his favorite pastimes, Einstein would often get lost. And he seemed to do it intentionally. Einstein's wife and friends said that's when he was the happiest—alone, adrift, and free to think. You can imagine he would likely not be too keen on the invention he helped birth: I assume the last thing he'd want is to be floating aimlessly in a lake and suddenly be interrupted by a disembodied voice telling him to turn right at the next dock.

And perhaps therein lies a lesson for all of us. Yes, GPS has enabled Chinese food deliveries to our doorstep. It means I might be your driver the next time you hail an Uber. It allows planes to fly and UPS to find my home in the woods. It makes it possible for us to go virtually anywhere we want, quite literally expanding our world. But sometimes, it's best to be lost and at peace. In a letter to Queen Elisabeth of Belgium, Einstein told her majesty about his love of sailing. "I have a compass that shines in the dark, like a serious seafarer," he wrote. "But I am not so talented in this art, and I am satisfied if I can manage to get myself off the sandbanks on which I become lodged."

In other words: sometimes it takes getting a little lost to find where you need to go.

EINSTEIN AND E.T.

"Why should Earth be the only
planet supporting human life?"

—ALBERT EINSTEIN

The search for alien life starts in the middle of nowhere, in a tiny town in West Virginia.

In his famous song "Take Me Home, Country Roads," an anthem here in Appalachia, John Denver proclaims the state "almost heaven." You can inhale the song's lyrics as you weave through the deep gorges, wide meadows, and craggy mountain byways. The state is utterly bountiful—in nature, at least.

Ranking number thirty-eight in state population with only 1.8 million people, it's tough to locate the "middle of nowhere" in a state as sparse as West Virginia, but it's possible if you try hard enough. Wind your way through the serpentine mountains— past the world's largest teapot and a life-size replica of Noah's Ark, past the deserted coal mining towns of Buckhannon and Mill Creek, past America's smallest post office, which just so happens to share a parking lot with America's smallest church—go past all of that and you're still not where this story begins.

But you're close.

In a valley between the Monongahela National Forest and the Allegheny Mountains lies the small town of Green Bank, West Virginia—population seventy-four. I can pretty much guarantee that the town of Green Bank is less inhabited than wherever you are right now, unless you're reading this from the remote island where the latest season of *Survivor* is being filmed. And even then—with all the camera crew, local guides, caterers, medical staff, and producers—there might still be more people on whatever South Pacific archipelago you're residing on. The town's most famous export was Bruce Bosley, a native son who went on to become a four-time all-pro with the San Francisco 49ers. He was a star running back at Green Bank High School, which has since been closed and demolished. But what's still standing, at 485 feet tall, is something extraordinary: the Green Bank Telescope and Observatory.

What makes the town unique and the reason that the telescope is there is that it's in the heart of the United States National Radio Quiet Zone. The "Quiet Zone" sounds like the fevered dream of a librarian at the Smithsonian, but it's real. A 13,000-square-mile area where radio transmissions are legally restricted for reasons of scientific and military research. But the real dead zone is the twenty miles surrounding the Green Bank Observatory, where the federal government forbids any radio transmissions at all. Police monitor the area for micro-wave ovens, Bluetooth signals, and Wi-Fi routers. Try to make a phone call, as I did while stepping out of the car, and you quickly realize there's no cell service. Finding a cell signal in

Green Bank, West Virginia, is just about the most difficult thing a human being can do.

Can you hear me now? No? Oh, you must be in Green Bank.

Even cordless phones are taboo here. It's like stepping back in time, to an era before TikTok and Uber and Alyssa Milano's Twitter feed. Inside the observatory, on a Friday afternoon, I met an elementary school teacher who told me that the kids in Green Bank enjoy a simpler life, one filled less with Netflix queues and more with outdoor activities. (Like figuring out which corner of town might have a working cell signal.) Norman Rockwell would feel right at home here.

For most, these restrictions are a nuisance. Trying to share a photo of Green Bank to your Instagram feed is a fool's errand. But a few dozen people have moved to Green Bank specifically because of the silence and a life free of iPhones. They suffer from electromagnetic hypersensitivity, a disease in which those frequencies can trigger acute symptoms like dizziness, nausea, rashes, irregular heartbeat, fatigue, and chest pains. They are the people who say using a cell phone gives them a headache, or that standing next to a microwave causes cancer. They are the electrosensitive. Some, I suspect, are merely refugees from the modern world, looking for serenity in a town that's completely devoid of all things Kardashian.

But this town is really a mecca for a certain kind of scientist.

The peace and solitude are what make this valley in central West Virginia just about the optimal spot to set up a telescope to search for aliens. Scientists looking to hear radio frequencies from outer space—especially those coming from beyond our galaxy—need perfect silence to do their listening. An alien spaceship may be trying to communicate with us, but it'll be hard to pick up what the little green beings are saying if invisible sound waves from cell towers and the *pop pop pop* of Orville Redenbacher's microwave popcorn gets in the way. The Radio Quiet Zone allows for the detection of faint radio frequency signals that man-made signals might otherwise mask. And that's how the world's largest fully steerable radio telescope ended up here—in the middle of nowhere.

The first thing you need to know about the Robert C. Byrd Green Bank Telescope is that it's massive. Imagine the biggest telescope you can and then double the size. No, triple it. It's the size of two football fields. The dish is so big that it's been called a "washbasin for Godzilla." It is the largest moving object ever built on land. It weighs seventeen million pounds and is about 60 percent taller than the Statue of Liberty. You can see it for miles. The locals refer to the GBT by its other acronym—the "Great Big Thing."

Like many points of interest in West Virginia—highways, museums, schools—it's named after the senator who served this state for more than half a century. Indeed, at the time of his death, Byrd was the longest-serving member in the history of the United States Congress. He was a controversial figure, but the state benefited immensely from the pork he brought home. Byrd parlayed his senior role in the Senate to push for funding of the telescope, which he received in the early 1990s. It took a decade to construct, and finally opened for scientific exploration in 2001. A year later, astronomers had already discovered three new neutron stars.

One particular scientist, Dr. Abraham "Avi" Loeb, the longest-serving chair of Harvard's esteemed astronomy department, has found particular interest in this area of Appalachia. The Israel-born astrophysicist who now calls Boston home is leading an international team of researchers in a search for intelligent alien life on another planet. So, it's no wonder that when he and his colleagues hatched an audacious plan to listen for radio signals from aliens, they descended on Green Bank, West Virginia.

As far as we know, Albert Einstein never stepped foot in West Virginia. He never partook in the moonshine famous in this area's backwoods. Despite his voracious appetite for unhealthy food, he likely never chowed down on a pepperoni roll, the state's distinctive and ubiquitous snack. But his influence and constant curiosity— *Are we alone in this universe?*—is felt every day at the Green Bank Telescope. Like many of the scientists who now conduct research here, the Nobel Prize–winning physicist had an affinity for the Red Planet. "It is a pity that we do not live on Mars," Einstein wrote a friend, "and just observe man's nasty antics by telescope."

It was the second part of that quote—the telescope—where most of Einstein's work was done. Much of his research did not take place in a laboratory, amid beakers and oxygen tanks, but in his famous thought experiments that focused on the sky above—Earth's gravitational pull, notions of time and space, the galaxy's black holes. Indeed, those astronomers who are looking for E.T. from a field in West Virginia are searching inside 100 billion galaxies. That needle in a haystack search comes courtesy of our friend Albert.

When Einstein looked up into the night sky a century ago, the known universe consisted of only a single galaxy. And in that world, in the early part of 1919, very few people outside of Germany had heard of Albert Einstein. Even Einstein himself was doubting his worth. He had just signed his divorce papers, ending a tempestuous twenty-year relationship with fellow scientist Mileva Marić. He was dealing with health issues. But then came May 29, 1919. That's the day when Albert Einstein became immortal.

Einstein had published his general theory of relativity a few years earlier, but it was still unproven. For the previous two centuries, Isaac Newton's logic had prevailed. The eighteenth-century British physicist believed that the galaxy had a static and fixed framework—and the fundamentals were unchanging. Einstein rewrote those rules throughout the early 1900s, stating that matter and energy could distort the universe in much the same way a person can cause a mattress to sag. And when that person rolls over, they would cause the mattress to shake and jiggle. Those ripples are called gravitational waves. If Einstein's predictions were proved accurate, it would mean recalculating where every star and planet is in the universe.

But there was a practical problem with Einstein's theory of relativity: it was just that, a theory. He couldn't properly measure his prediction without seeing those gravitational waves in motion, and telescopes in the early 1900s could not detect such subtlety in space.

But what if there was a moment when such movements were not so subtle?

What if Einstein could know in advance when something so big was about to happen that the astronomers of his day could spot it from Earth? As it happened, Einstein was in luck. A solar eclipse was scheduled for the spring of 1919, providing the physicist the perfect opportunity to test his theory. During a total solar eclipse, stars passing through the sun's gravitational field would be visible, and accurate measurements could, in fact, be made. It's kind of like how Marty McFly in *Back to the Future* knew in advance when and where a bolt of lightning was about to strike.

All Einstein had to do now was figure out a way to photograph an eclipse.

So, he enlisted the help of Sir Arthur Eddington, a British astronomer. Eddington traveled to the island of Principe off the western coast of Africa to take pictures. They also sent astronomers to Sobral, Brazil, just in case it was raining in Africa and they needed backup photos. (Indeed, an attempt by Einstein to photograph an eclipse the previous year was marred by cloudy weather.) And then it happened. For six minutes and fifty-one seconds on May 29, 1919, Einstein and his team witnessed one of the longest solar eclipses of the twentieth century.

They analyzed the data from the photographs and held a press conference to announce their findings at London's Royal Society, the oldest national scientific institution in the world (and which counts Isaac Newton as an early member). In that one moment, the theory of relativity overthrew Newton's law of gravity as the reigning rule in physics. The *New York Times*, which was so unprepared for the news that it sent its golf reporter to cover the announcement, declared in an all-caps headline: "EINSTEIN THEORY TRIUMPHS" and went on to call it "perhaps the greatest of achievements in the history of human thought." Up until that point, we assumed that if we went in a straight line, we could travel indefinitely through space. What Einstein found was something entirely new: Space is curved. Travel far enough and you eventually end right back where you started. It's kind of like Christopher Columbus discovering the Earth wasn't flat.

Einstein instantly became famous, recognized wherever he went. He received so much fan mail and press inquiries that he complained he could "barely breathe." It was the Big Bang moment of Einstein's life. Two years later, he was awarded the Nobel Prize in Physics.

"It's hard to think of a more important experiment in the twentieth century," Daniel Kennefick, the author of a book about the historical eclipse of 1919, told me when I called him up. The scientists on the expedition overcame war, bad weather, and equipment problems to test and prove one of science's now most fundamental suppositions—Einstein's theory of relativity. (In 2008, HBO dramatized that solar eclipse in a movie called *Einstein and Eddington* starring David Tennant from *Doctor Who* and Andy Serkis from the Lord of the Rings trilogy as the two scientists.)

That look into the abyss, for celestial proof of a scientific concept, was aided by the use of the most powerful telescopes of the day. A century later, Einstein's scientific descendants—like

Avi Loeb and his fellow alien hunters in West Virginia—are using one of the most technologically advanced telescopes to find something else in the vast sky above.

Today's particular search for aliens, for evidence of intelligent life beyond our human race, may be happening in a field in West Virginia, but the mission's pedigree is far from backwater. Facebook founder Mark Zuckerberg is on the board, as was physicist Stephen Hawking. (This was one of the physicist's final projects before he passed away in 2018.)

With pomp and circumstance, the project was unveiled on July 20, 2015, the anniversary of the Apollo 11 moon landing. Almost one hundred years after Einstein's press conference, Hawking took to the same stage at London's Royal Society to make a historic declaration: "In an infinite universe, there must be other life," Hawking said at the public launch. "There is no bigger question. It is time to commit to finding the answer."

The ambitious mission was christened "Breakthrough Listen"; its sole purpose would be to search for intelligent communications from outside our solar system. And it would be paid for by an unlikely source: an Israeli-Russian billionaire. Yuri Milner, a venture capitalist and a physicist himself, has been called one of the world's greatest leaders by *Fortune* magazine, and *Time* named him one of the one hundred most influential people in the world. He was an early investor in Facebook, Twitter, WhatsApp, Snapchat, Airbnb, and Spotify. And he's behind the world's largest scientific award, known as the Breakthrough Prize. Each winner receives $3 million in prize money. Milner—who was named after Yuri Gagarin, the first human to travel into outer space—launched the Breakthrough Listen initiative with a personal gift of $100 million.

The world watched, with collective mouths agape, as the announcement was made. One hundred million dollars could certainly go a long way to solving more pressing matters—like curing cancer or stopping telemarketers. But Milner had some insight that the rest of us were not privy to. He knew this wasn't a total waste of his money, and he had been convinced of this by a man who calls Einstein his intellectual hero—Harvard's Avi Loeb.

Loeb received a bachelor's, master's, and PhD in theoretical physics from the Hebrew University in Jerusalem—a school that Einstein helped establish on a trip to Palestine in 1923. Loeb then followed Einstein's path to the Institute for Advanced Study at Princeton University. After spending five years there, he was eventually hired by Harvard University, where he quickly rose up the ranks—starting first as an assistant professor and getting tenure after only three years. His work is prolific, publishing eight hundred scientific papers on everything from exoplanets to black holes. *Time* magazine, the same one that honored Einstein as the Person of the Century, named Loeb one of the twenty-five most influential people in space exploration. He soon took over Harvard's astronomy department.

As the guy who manages Albert Einstein's social media accounts, I'm constantly on the lookout for intriguing news nuggets to share with the genius's twenty million fans (like the hard-hitting story about a parrot at the Knoxville Zoo who has such a vast vocabulary that they call the bird Einstein), so I had come across Dr. Loeb's name often. In a press release about gravitational waves, there was a quote from Dr. Loeb. In another, about a scientific finding related to black holes, he was pontificating on the meaning of the discovery. He was the Zelig of Einstein headlines.

An article I posted about how Einstein's theories were helping alien hunters like Loeb was one of the most popular I'd ever posted on Facebook, garnering more than seven thousand "likes" in just a few hours. It was shared by thousands more. The fans loved it. One commentator noted the scene in the 1977 alien film *Close Encounters of the Third Kind* where a scientist said of the extraterrestrials: "Einstein was probably one of them." Another comment was just a non-stop GIF of two aliens giving each other a high-five.

That's when I knew I had to visit Loeb in person. I had struck a nerve. So much of what I shared with Einstein's adoring public was ephemeral (again, allow me to remind you of Einstein the parrot). But posting about Loeb's alien quest had tapped into something else entirely. He was continuing Einstein's research and applying it to a field that everyone—from armchair physicists to online trolls—found fascinating. His work was, by definition, keeping Einstein's legacy alive and relevant to a modern-day audience. And I wanted to meet the man who, without even having his own social media account, had somehow cracked the code of what goes viral online. So, I booked a ticket to Boston.

Dr. Loeb's office, much like an alien planet, is a little hard to find. The Center for Astrophysics on the campus of Harvard University is actually a complex of buildings. Finding the right one, then the right floor, then the right hallway, should've been relatively easy—except that it's Presidents' Day, so the school is closed and most of the buildings are locked. Thanks, George Washington. Your birthday may be a boon to furniture sales, but 20 percent off an ottoman was not going to help me interview the world's leading extraterrestrial expert.

I called Loeb's cell phone when I arrived, and he met me outside in the parking lot to let me in. He comes across as bookish; a tense grin belies a bewilderment that his research has caused such a media maelstrom. I explained to him the basic rules of journalism, that when the chair of Harvard's prestigious astronomy department credibly talks about aliens, this is beyond the domain of the *National Enquirer*. It's no wonder the *New York Times*, the *Wall Street Journal*, and CBS News have all done features about him.

But he wasn't always destined for greatness. "I'm doing astronomy, but this is not really my first love," he confided to me in his Mediterranean accent as we walk up the staircase to his second-floor office.

Loeb's grandparents fled Nazi Germany before World War II and emigrated to the bucolic village of Beit Hanan in Israel. Loeb was born in the winter of 1962 and was raised in the agrarian ways of his parents. His mom was a baker, and his dad was Israel's leading pecan farmer. "I grew up on a farm, and I always had the option of staying on the farm rather than being in academia," Loeb said. "As a young kid, I used to collect eggs every afternoon and drive on the tractor through the hills that my family owned and read philosophy books. I dreamed of thinking for a living." When Loeb was a teenager, his mother started studying for a PhD, and he would sometimes sit in on her philosophy classes. Like Einstein, he said, his mind worked in thought experiments, which gave him "the ability to see things from a distance and to look at the big picture."

We arrived at his office, a typical bland affair complete with beige filing cabinets and a metal desk that looks like it came from Office Depot, which completely disguised the fascinating research being done by its inhabitant. Loeb himself—meek, thin, and dressed in all black—looked like someone who could use a media makeover. He looked like a tenured astrophysics professor, but at the same time, he considers himself a rebel within academia—someone who shouldn't have to color inside the lines.

After high school, Loeb was drafted into the Israeli military's elite Talpiot program, for those who showed outstanding academic ability in the sciences. Instead of spending time on the battlefield, they are tasked with coming up with technological innovations. In the early 1980s, a twenty-one-year-old Loeb figured out a way to make missiles travel ten times faster than existing speeds by employing a simple switch in energy. He presented his findings to a visiting dignitary from the United States who turned out to be in charge of Ronald Reagan's Star Wars military defense program. Reagan's administration agreed to fund Loeb's project and put him in charge of thirty scientists to make it happen.

Loeb told me that if he hadn't gone into the military, he would've become a philosopher. "And nobody aside from philosophers would know about me," he said with a smile. "I'm a rare bird."

He handed me a coffee mug with the logo of the Black Hole Initiative emblazoned on it. Loeb is the founding director of the group. "It brings together astronomers, physicists, mathematicians, and philosophers," Loeb explained. "And on a daily basis, we basically discuss puzzles and interesting questions that are unsolved at the moment about black holes."

In 2009, Loeb and one of his postdoctoral fellows, Avery Broderick, were the first to publish a peer-reviewed paper predicting what a black hole would look like. A decade later, in April 2019, Loeb and his team of astronomers gained international notoriety when they held six simultaneous press conferences across the globe in Belgium, Chile, Shanghai, Japan, Taipei, and the United States. At the event, they revealed something no human had ever seen before: the first-ever photograph of a black hole.

Black holes were first described a century ago as a mathematical concept from Einstein, a cosmic abyss so deep and dense that not even light could escape it. A black hole is a celestial object that compresses a huge mass into an extremely small space. According to Einstein's theory of relativity that he proved during the 1919 eclipse, light should bend around a black hole, creating a circular silhouette.

Ask a physicist what photographing one would entail, and you're likely to hear a list of superlatives: "That's not an impressive feat. That's a mind-blowing feat. It's a technical tour de force like we've never seen before," said Scott Hughes, a theoretical physicist at MIT. Columbia University physicist Brian Greene explained that photographing a black hole would be "one of the most thrilling discoveries of our age."

In other words, snapping a picture of a black hole is a lot tougher than taking a selfie. At twenty-six thousand light-years away, the closest one to Earth is incredibly far. It's so small that there's no telescope in the world that can spot it. It would be the equivalent of trying to see an avocado on the moon with the naked eye. So, the scientists—more than two hundred of them across sixty countries—channeled Einstein and thought outside the box, beyond the scope of their current knowledge. They built a fundamentally new instrument to detect something that small. It required devising a master plan. They combined a global network of telescopes in Spain, Mexico, Arizona, Hawaii, Chile, and the South Pole. Put together, they dubbed it the "Event Horizon Telescope." To capture the crucial image, all the telescopes had to point to the exact same spot at the exact same moment.

In the end, they were successful, revealing to the world a bright ring of light circling the shadow of a black hole. Einstein's original prediction of a porthole to eternity was now visible for all to see. As described by the *New York Times*, the image of the "black hole appears as a finely whiskered vortex, like the spinning fan blades of a jet engine, pumping matter into the black hole and energy outward into space." In the center is a black hole with the mass of six billion suns (that's like a 2 with 36 zeros after it) swallowing any light particles that stray too close. It is profound evidence that verified the existence of black holes first predicted by Einstein's theory of relativity.

"It's going to be the discovery of my lifetime," said Shep Doeleman, who works with Loeb at the Black Hole Initiative in Cambridge. "It's sobering to see what a black hole looks like for the first time." That image is now on display in the Museum of Modern Art in New York. Two years later, in the spring of 2022, scientists had already photographed another.

For Loeb and his postdoc Broderick, it was the proof that their initial prediction a decade earlier was correct. "Sometimes the math looks ugly, but there's really a strong aesthetic in theoretical physics. And the Einstein equations are beautiful," Broderick said. "In my experience, nature wants to be beautiful, and that's one of the striking elements of Einstein's description of gravity. It is fundamentally one of the most beautiful theories that we have. For that reason alone, and the long history of Einstein being proven right, I suppose we're not terribly surprised."

It was as if Loeb was reaching back through time and grabbing the baton from Einstein himself. "Without this understanding that Einstein provided, we would never get to that point," he said.

Within a decade, the team hopes to produce the first-ever moving image of a black hole. Videos can help them drill into the details of how black holes consume matter and affect the galaxies they find themselves within. What's more, these black hole videos could one day help scientists push the bounds of physics as we know it, potentially finding new tests for Einstein's theory of relativity.

Like Einstein, Loeb had used the power of the telescope to shine a light on a mystery from deep space. For his next discovery, Loeb would once again turn to the telescope, to look beyond black holes and even further out into space. And that's where the aliens enter our story.

Nearly 3,000 miles away from Loeb's Harvard office is ground zero for America's love affair with aliens. That's where you'll find the Area 51 Alien Travel Center, a combination diner, convenience store, and bordello in Amargosa Valley, Nevada, about ninety miles northwest of Las Vegas. The place has a surprisingly decent TripAdvisor rating, considering it somehow manages to be, according to one reviewer, a "nice place to buy some attractive presents for my grandchildren" while at the same time, according to another, being home to "America's only themed brothel," complete with an Alien Probe room.

Quirky roadside attractions sprinkle the nearby ninety-eight-mile-long "Extraterrestrial Highway" that winds through this section of the Nevada desert, which looks, appropriately enough, like a Martian landscape. A sign for E.T. Fresh Jerky offers passersby an opportunity to "drop your toxic waste in the cleanest restrooms in Area 51." A kitschy café and lodge—cleverly called the Little A'Le'Inn—advertises its policy of "E.T.s and Earthlings Welcome Always." It's no surprise that the route has become a popular tourist destination.

In the summer of 2019, two million people joined a Facebook group promoting an alien-themed event in the area. Humorously dubbed "Storm Area 51, They Can't Stop All of Us," the idea was the brainchild of a college student named Matty Roberts, who meant it as a joke, but who clearly didn't expect it to go viral. "So, the FBI showed up at 10:00 a.m. and contacted my mom and she calls me like, 'Answer your phone. The FBI is here,'" Roberts told his local ABC affiliate. "I was kind of scared at this point, but they were super cool and wanted to make sure that I wasn't an actual terrorist making pipe bombs in the living room." Roberts himself eventually pulled out of his own event, afraid it would become another Fyre Festival, the notorious ill-planned music gathering that was supposed to be held on one of Pablo Escobar's islands in the Bahamas. "I didn't feel comfortable with inviting even my friends and family out to this event, let alone thousands of strangers," Roberts told the *Washington Post*, going so far as calling his event a potential "humanitarian disaster."

And yet, our collective fascination with aliens is all too real. So, the event went on without Roberts. On Einstein's Facebook page, I shared an article about a new study from a group of physicists at Penn State University. The scientists claimed that extraterrestrials had likely already visited Earth billions of years ago. And what's the reason they haven't returned? Well, apparently, humans bored them. The story was viewed by a quarter million people in just twenty-four hours.

For years, pop culture has been telling us that aliens exist—over and over again. (Did we really need a fourth *Men in Black* movie?) But as the slogan for *The X-Files* warned us: The truth is out there. And that truth may lie in the Nevada desert. Located just outside Death Valley National Park, the airbase known as Area 51 has long been the subject of conspiracy theories, with many speculating that it could be the government's hiding place for secrets about aliens and UFOs.

Since 1989, Bob Lazar has been telling anyone who will listen that he used to work at Area 51. Indeed, he claimed that he'd once worked on a recovered alien spacecraft for the U.S. government. His job, he said, was to reverse-engineer a material called "Element 115," which was apparently what the aliens used to gas up their spaceship. "It's a super heavy element," Lazar once told Larry King. "It's a unique element. When it's exposed to radiation, it produces its own gravitational field—its own anti-gravitational field—and it's what's used to lift and propel the craft." The FBI had enough of Lazar's antics that, in 2016, they raided his warehouse where he kept his souvenirs from his work at Area 51—although they claimed they were looking for a poisonous gas that was used in a Michigan murder.

A quick Google search and you'll soon discover that Bob Lazar is just the tip of the quack pot iceberg when it comes to Area 51. The internet is a proverbial rabbit hole of hypotheses about the place: It's where military officials have secret meetings with extraterrestrials. It's the United Nations of alien civilizations. Some even believe it's where scientists are working on teleportation technology and time travel, a topic that was of great interest to Einstein.

Our modern-day fascination with all things alien is a perfect barometer for what can happen when the naïve and gullible meet science they don't quite understand. The 1938 radio broadcast of *The War of the Worlds*—an early precursor of today's fake news—caused a nationwide panic. But as Einstein reached his twilight, it was the advent of the motion picture that brought with it more vivid imagery of what we believed aliens would look like. A through-line can be drawn from the B-movies of the 1950s and 1960s to the ones that dot our multiplex today. With names like the *Invasion of the Saucer Men* and *The Man from Planet X*, those early entrants might lead one to think that there was no such thing as aliens of the fairer sex. But the 1984 cult TV series *V*, about lizard-like aliens who take the form of human bodies, was sure to change those perceptions. (As a nine-year-old-boy, I remember being oddly attracted to the Supreme Commander Diana.)

Ridley Scott's *Alien* kicked off a franchise that has spanned decades. Tim Burton's *Mars Attacks!* showed us that we needn't take every alien invasion so seriously. And things came full circle when Steven Spielberg and Tom Cruise took on *War of the Worlds* themselves in 2005. By then, nobody mistook Cruise battling aliens—or, for that matter, believably portraying a down-on-his-luck dock worker—as the real deal.

I asked Loeb what kind of space movies he enjoys. "I have to confess that I don't like science fiction," he replied, laughing. Like Einstein, Loeb is not a fan of speculative fiction. "I get upset about things that violate the laws of physics, so I cannot enjoy so much the artistic aspects of the movie, when I see things that don't make sense. It bothers me. An unrealistic script resembles, in my mind, a Ponzi scheme where the illusion may be beautiful but misleading."

Instead of being a fan of the Star Wars and Star Trek franchises, his favorite space movie in recent years is *Arrival*, based on a short story by Ted Chiang called "Story of Your Life." It stars Amy Adams as a language expert enlisted by the U.S. Army to discover how to communicate with aliens who have arrived on Earth before tensions can mount between the two species. "I loved this film because it addresses the most important challenge when encountering an alien civilization: communication. The difficulties in interpreting signals from an alien culture could be far greater than those faced by Alan Turing when cracking the Enigma code of the Germans during the second World War," Loeb explained. "My prediction is that there will be a new research field of 'astro-linguistics' in the future."

He cited the 2015 movie *The Martian*, adapted from the novel by Andy Weir, as another film that basically stuck to the facts. It starred Matt Damon as an astronaut abandoned on the Red Planet who has to figure out creative, but fact-based, ways to survive. "It has an excellent depiction of the physical challenges for humans living on Mars," Loeb said, stretching back in his chair, clearly enjoying this intellectual exercise. "The film is as close to being realistic as science fiction gets." Loeb was particularly pleased that NASA helped director Ridley Scott with the making of the movie. The space organization not only answered questions about rockets and rovers, but also invited the filmmaking crew to see prototypes of actual Mars habitats that NASA was in the process of building.

"The thoughts about outer space are often a reflection on our inner space," Loeb said. "We learn about ourselves from imagining what may be lurking out there. This is the beauty of science fiction."

When Albert Einstein was growing up in Europe, he was fascinated with the sky above. A love of science fiction is what he credited for inspiring him to think about space travel. In particular, he was a fan of author Aaron Bernstein, a fellow German, whose seminal 1880 work *Popular Books on Natural Science* encouraged the young Einstein to pursue physics. (In fact, when the book was re-published in 1923, Einstein wrote the preface.) One of Bernstein's stories—which imagined an electric pulse coursing through a telegraph wire—is said to have inspired Einstein to devise his famous thought experiment of traveling along with a beam of light, which eventually led to his theory of special relativity.

By the time Einstein was a teenager, in the 1890s, his mind was already well beyond his years. At the age of twelve, over the course of a single summer, he taught himself algebra and Euclidean geometry. He was reading books on philosophy by Immanuel Kant. In 1895, when Einstein was first forming his ideas about physics and space, H. G. Wells's classic book *The Time Machine* was released. While most believed the known universe existed in three dimensions (height, length, and width), Wells posited a fourth: time. The idea intrigued the young Einstein, whose theories would later blend the concepts of space and time.

Einstein and Wells would eventually meet in 1929 in

Berlin, where the science fiction author was speaking about a passion he shared with the physicist: world peace. A photograph from the event, where Einstein was an honored guest, shows the men seated next to each other. After World War I, both men were horrified with twentieth-century warfare and devoted much of their time to promoting international cooperation and pacifism.

Einstein was often displeased when he encountered his fellow humans—whether it was their obsession with war or, in moments of mundanity, the social graces of everyday life. He didn't understand the need to brush his hair or wear socks. Instead, he preferred to cohabitate with the stars above, their twinkling lights offering solace from society, along with more universal truths. Astronomer Edwin Hubble, known as "the Columbus of the skies" for proving that millions of galaxies exist beyond our own, once shared an awe-inspiring moment with Einstein in 1930 at the Mount Wilson Observatory in California. "Hubble gave Einstein a turn at the 2.5-meter snout of the world's largest telescope to peer into the heavens he so often contemplated," wrote Carolyn Abraham in *Possessing Genius*. "Einstein glimpsed the galaxy-studded cosmos for the first time and knew that Hubble was right: humanity occupies a dot in an ever-expanding universe." It was during these times when Einstein pondered if perhaps there was another place, with another life-form, that would be more to his liking than his home planet. "There is every reason to believe that Mars and other planets are inhabited," Einstein told a reporter. "Why should the Earth be the only planet supporting human life?" In the spirit of world peace, you'd have to wonder what message Einstein would send out to an alien civilization.

Fellow physicists were given just such an opportunity not long after Einstein's death. In the fall of 1961, a group of researchers—including a young Carl Sagan—traveled to Green Bank, West Virginia, for a conclave on the topic. Calling themselves the "Order of the Dolphin," they began devising a way to communicate with aliens.

By the fall of 1974, technology caught up with their ambitions. Astronomers built the Arecibo Observatory, which was, at the time, the world's largest radio telescope. Located in the jungle of Puerto Rico, its nine-hundred-ton platform the size of three football fields was suspended above the trees. And they beamed a three-minute audio message toward a narrow cluster of three hundred thousand stars.

NASA was emboldened by this effort and, in 1977, came up with a plan to slingshot a pair of spacecrafts around a succession of planets and off into outer space. They wanted to send a message to any potential alien civilizations that Earth was inhabited by a peace-loving community. With only six weeks to compose a message, Sagan and his team spent hours coming up with the perfect time capsule. They ended up including images, anatomical drawings, astronomical maps, technical diagrams, and pictures.

They also included a lot of sounds. Songs of humpback whales, crickets and frogs, brainwaves and spoken greetings in fifty-nine different human languages. The pièce de resistance was a vinyl record; Sagan figured that surely the extraterrestrials would be technologically advanced enough to listen to an LP. Encoded on it were thirty-one tracks that included songs as varied as Beethoven's "Opus 130" to Chuck Berry's "Johnny B. Goode." Plated in gold, it was constructed to survive for a billion years. Who knows? The enchanting artifact may be the only thing left of our civilization once we're gone. In the meantime, the musical playlist of the Voyager Golden Record, as it was called, has since been burned onto a CD and is available for sale on Amazon for a whopping $225.

Back on the island of Puerto Rico, the Arecibo Observatory would eventually die. After decades of use, it literally began falling from above the tree line and, one morning in December 2020, it came crashing down in the jungle.

But remember that message that astronomers beamed out in space back in 1974? They didn't just shoot it into nowhere, they aimed it at a star cluster called M-13. And that star cluster is twenty-five thousand light-years away. So that original Arecibo message is still out there zooming through the cosmos at the speed of light. It's been traveling for about half a century so far. It's still got more than twenty-four thousand light-years to go. If and when that message ever reaches that faraway star cluster, an alien may manage to decode it. Even though the telescope itself is long gone, in many ways, its significance transcends its utility.

Sadly, Einstein was not able to see any of humanity's forays beyond our planet. He died at seventy-six years old, shortly after midnight on April 18, 1955; had he lived another two years, he would have witnessed the Soviet Union sending its first *Sputnik* rocket into Earth's orbit, officially launching the space race. Another seven years and he would have heard John F. Kennedy's 1962 "We choose to go to the moon" speech. A few years more and he could have watched with awe as the images of Neil Armstrong and Buzz Aldrin stepping foot on the moon were beamed back down for all to see on television. Years later, back on terra firma, Aldrin credited Einstein with inspiring him. "My philosophy is more like what Albert Einstein called a cosmic sense of a greater power involved in the creation of the universe," Aldrin told *National Geographic* in 2016. It's almost certain the great physicist was prepared for others to continue his search for science's answers. "Our death is not an end if we have lived on in our children and the younger generation—for they are us," Einstein once told a grieving widow. "Our bodies are only wilted leaves on the tree of life."

Seven years after Einstein passed away at a hospital in Princeton, a boy was born in a small village in central Israel. His parents, farmers whose vision of a black hole was likely something earthlier, bestowed the baby with a propitious name. "We share the same first name," Loeb told me at his office in Harvard. "I'm called after my grandfather, whose name was Albert. Abraham is the Hebrew name for Albert."

The search for intelligent alien life began at a goat farm in Israel. Loeb remembered the day vividly. He was on vacation in the mountains of Israel when he got the call. Milner, the billionaire, was asking him a seemingly impossible question: Could Loeb come up with a plan to find aliens, and could he get a presentation ready in a matter of days?

Never one to shy away from a challenge, Loeb knew he needed to get online fast. The only place he could get connected to the internet was in the ramshackle office at the goat farm where they had a computer at the reception desk. The irony was not lost on him: here he was, the man behind a bold attempt to meet aliens, and he was sitting in the dirt staring at goats.

"I basically sat at 6:00 a.m. working on this presentation, with my back to the wall of this office, looking at the goats that were just born the day before, and contemplating the first realistic plan to send a spacecraft to the nearest star," he told me. "And I'm sure that the owner of that goat farm never imagined that this would happen."

A week later, Loeb was at Milner's mansion, PowerPoint in hand. Loeb's slides showed that it was possible not only to find aliens, but also to indeed find a new planet for human civilization. The billionaire was impressed and, shortly thereafter, announced the Breakthrough Listen initiative. He made Loeb the chairman of the project's advisory committee. Together, they became the Lewis and Clark of interstellar travel.

"We reached a conclusion that there is only one technology at the moment that looks feasible, in principle, that doesn't violate the laws of physics," Loeb said. Specifically, this is their challenge: They want to send spacecrafts to the Alpha Centauri star system, the closest planetary system to our own. That's about twenty-four trillion miles away, and it would take us about twenty years to get there. As if that weren't enough, it's four light-years away. Einstein taught that you can't travel faster than the speed of light. So, notifying Earth of a successful arrival would be like finding out Donald Trump won the 2016 presidential election after Joe Biden had already taken office.

Loeb discovered a cost-efficient way to make this happen. It involved using a powerful laser beam—one Einstein first described in a 1917 paper—to push tiny spaceships outside our solar system. It may sound implausible, or something out of an Austin Powers movie, but the initiative has already seen its first tangible success. In 2017, they launched the world's smallest spacecraft. How small were they? Called "Sprites" (no, not the soda in the green can), these 3.5-by-3.5-centimeter chips weigh just four grams but contain solar panels, computers, cameras, and radios on a single tiny circuit board. They're about the size of a Ritz cracker.

The next step is developing an array of medium-scale lasers back on Earth that will combine to form one large, powerful beam that can push the tiny spaceships, which are attached to a lightweight sail. Think of this push like air from a fan, propelling forward a fleet of a thousand miniature satellites. Once these spaceships arrive, they can take photos and report back to Earth what they find.

In 1935, Einstein told an aspiring teenage astronomer that he believed there was a "natural presumption that life in some form may not be unique to our planet." Loeb's search for alien life takes Einstein's presumption to the next level. It's less a theoretical, cosmic quest to answer an age-old question: Are we alone? For Loeb, it is more existential. He knows that our time on Earth is finite. Climate change is wreaking havoc on our planet. Someday, he noted, the sun will eventually boil the oceans to the point where we'll be forced to find a new home. Even beyond the effects of global warming, there could be a nuclear war or a catastrophic asteroid impact. None of this will happen tomorrow, or perhaps not even in the next century. But the mild-mannered Loeb knows it will happen at some point. "People may prefer to ignore it. You could ignore any danger in the future, but eventually it will come to haunt you because nature does not care about what we think and what we ignore. The reality will come to haunt us."

That is why finding a new Earth is on his to-do list. He wants to explore the universe for a planet that is habitable to life and ask the presumably peaceful aliens if we can join them and move in. Think of a group of humans boarding a galactic Noah's Ark to restart humanity on another planet. Livable planets just outside our solar system have already been discovered. At the top of his list is one dubbed "Proxima B."

"This planet would be an excellent target to visit," he said. "That's a good neighborhood. In fact, I'm trying to convince my wealthy friends to invest in real estate on the planet." Loeb explained that the planet is locked to a nearby star in such a way that it has a permanent day side and a permanent night side, and the most valuable plot of land may be right in between those two. "I think the highest value for real estate would be on this strip that separates the two because you can witness, through your windows, a permanent sunset."

Loeb is figuring out how to get us there. "We just need to think about the big picture and, you know, have a plan B."

It's possible that aliens are already living there, which is why Loeb is so eager to find them. On the morning of October 19, 2017, just two years after Breakthrough Listen launched, Loeb's research got an unexpected jolt. The massive Green Bank Telescope in West Virginia and another at an observatory in Hawaii spotted something strange in the sky above. It was the size of a football field. The scientists dubbed it "Oumuamua," the Hawaiian word for "scout," and it's believed to be the first interstellar object detected passing through our solar system. In 2019, Loeb confirmed he had spotted a second.

In the summer of 2021, when the U.S. government released a highly anticipated report about UFOs, donors offered Loeb $1.7 million to launch a new initiative. Called the Galileo Project, it's named after the seventeenth-century Italian scientist who, like Einstein, enjoyed challenging the establishment. It would use the Green Bank Telescope and other large telescopes around the globe—like the Vera C. Rubin Observatory in Chile—to monitor the skies for evidence of intelligent alien life. That same year, Loeb published a book, *Extraterrestrial*, which skyrocketed up the *New York Times* bestseller list.

Ask Loeb what he thinks the strange flying object was and he'll tell you, in all earnestness, that it's possible it was an alien ship on a reconnaissance mission. To the surprise of few, his suppositions are not without controversy. As Carl Sagan once famously said: extraordinary claims require extraordinary evidence. Many scientists, even those who respect Loeb's work in general, believe the Harvard professor is too quick to always point to aliens. They think he's an interstellar Paul Revere. "The aliens are coming! The aliens are coming!" They say Oumuamua was likely just a comet or a chunk that had broken off another planet. "I welcome other proposals," Loeb said, "but I cannot think of another explanation for the peculiar acceleration of Oumuamua."

It's not dissimilar to the way the scientific community first treated Einstein and his "out there" theories. When Einstein redrew the fundamental rules of how our universe operates, he was initially viewed as a crackpot. A 1931 book titled *One Hundred Authors against Einstein* featured essays by dozens of scientists declaring the theory of general relativity a bunch of malarkey.

I asked Loeb if he sees parallels in how the establishment treated Einstein with how the scientific community often scoffed at his quest to find extraterrestrials. "Definitely," he said, without missing a beat, "because a lot of people try to establish authority. In science, authorities are not the relevant tools to advance knowledge. You see that in politics quite often that a lot of people agree on something. And it turns out to be wrong, and you see that in religion, you see that in many other aspects of life."

Einstein was an outsider, in terms of both his theories and his place in the world—Jewish at a time when it was dangerous to be so. Loeb, whose grandfather Albert lived in Germany at the same time as Einstein, tries to embody the powerful combination imbued by his intellectual hero—"chutzpah and courage," he explained, using the Yiddish term for confidence. Talking about Einstein's reimagining of Newton's law of gravity, Loeb's eyes lit up. "I cannot think of anything more brilliant," he said. "It's remarkable. It's the biggest intellectual achievement that I can see in the history of physics."

And yet, even Einstein wasn't always correct. He made several predictions in the final decade of his career that turned out not to be true. "When you work at the frontier, you don't always know what's right, what's wrong," Loeb said, echoing his own quixotic search for aliens. "But nowadays, you find this culture of scientists where they don't put skin in the game, they don't make predictions that can be proven wrong. That's the most comfortable position and then you can say that you're always right. You'll get the honors and awards. But what's the point?"

From the platform atop the Green Bank Telescope, Loeb is about as close to space as you can get in West Virginia. Already, he and his colleagues have collected two petabytes of data, which sounds like a lot because it is. It's about two million gigabytes of information. (This would be the equivalent of storing around 120,000 Nicholas Cage movies in HD.) They have observed more than 1,300 relatively nearby stars over the course of just a few years, listening for any signs of radio waves that would signal the presence of technologically advanced extraterrestrials. This only amounts to a tiny sliver of what could be studied. If you compare the volume of space we're able to search for signs of advanced technology to the volume of the Earth's oceans, then "so far since 1960, we've searched about one hot tub's worth of the ocean," explained Jill Tarter, a longtime alien researcher.

In total, Breakthrough Listen's goal is to survey one million stars and one hundred nearby galaxies for any kind of technology not placed there by humans, which would provide proof of the presence of alien life. The entire project will last a decade— unless Milner decides to toss in even more of his vast fortune. To keep the fast pace of discovery moving, the group publicly releases everything it finds. In June 2019, it dropped the largest data dump of its kind, allowing scientists from across the globe to parse the findings.

While the project may be audacious, Loeb sees humility as its guiding light. "When you enter a room full of strangers," he said referring to meeting aliens, "you better listen first before you speak out, because you never know whether there are predators out there."

Picture this, Loeb told me in his office. "We are born into this world like actors put on a stage without a script. The first thing to do is examine the stage, and that's the universe. It's much more than the Earth. If you look up in the sky, you realize that we are part of a much bigger stage—that's the cosmos. And Einstein provided the theoretical framework that allows us to understand the cosmos."

Loeb's search for extraterrestrials takes that analogy to its next inevitable conclusion. Are there other actors on this stage? "If there are, perhaps they know the answer of what the play is about," he said. "Maybe they will tell us something for which we don't have the script. Maybe they'll tell us something that we don't know, maybe they are wiser, because they've been around for billions of years more than us. So, we could learn from others, from smarter kids on our block."

Besides listening for aliens who are currently out there, Loeb is also searching for evidence of alien civilizations that may no longer be with us. He is fascinated by a growing field of research known as "space archaeology" and is actively looking for relics of the past floating in deep space. "Instead of using shovels to dig into the ground as is routine in conventional archaeology, this new frontier will be explored by using telescopes to survey the sky and dig into space," he said.

Meanwhile, other teams of scientists are sending messages into space. This high-tech "message in a bottle" proposal has plenty of critics—perhaps the aliens will misinterpret the message and attack Earth. "I think 99 percent of astronomers think that this is a bad idea," said Dan Werthimer, a scientist with Search for Extraterrestrial Intelligence, or SETI, Institute at the University of California, Berkeley. Even Stephen Hawking, a proponent of Breakthrough Listen, was wary of going too far, warning that aliens might be "looking to conquer and colonize" planets they visit.

Of course, it's also possible—perhaps likely—that scientists will find alien life in another way that has nothing to do with hunting for faint radio signals. Scientists hope to one day search water-rich moons like Enceladus (Saturn) or Europa (Jupiter) that might harbor microbial life. NASA's next-generation telescopes could give scientists a glimpse of distant planets that might have environments ripe for life. NASA's *Curiosity* rover has already sniffed out methane on Mars, which could be a tantalizing hint of microbial life up there. Call it an alien burp.

Ever the explorer, Loeb believes we should continue scanning the sky. "Such a search would resemble my favorite activity with my daughters when we vacation on a beach—namely, examining shells swept ashore from the ocean. Not all shells are the same, and similarly only a fraction of the interstellar objects might be technological debris of alien civilizations. But we should examine anything that enters the solar system from interstellar space in order to infer the true nature of Oumuamua or other objects of its mysterious population."

One of the tools at Loeb's disposal is something he pioneered for finding planets around other stars called "gravitational microlensing," otherwise known as nature's magnifying glass. It too is rooted in Einstein's theories and his eclipse work in 1919, as it helps scientists spot a new planet by noticing the way light bends around it. "About one hundred years ago, Einstein came up with his theory of gravity where gravity is not a force. It's just a distortion of space and time. And we use this insight in anything we do these days in astrophysics," Loeb said.

As with so many aspects of Loeb's research, Einstein serves as muse, the "ur-physicist" whose work remains relevant—more so today than ever before. His mere five-character mathematical equation, $E=mc^2$, transformed him into a galactic clairvoyant and perhaps the greatest scientist who ever lived. And it's that theory that continues to bear fruit for scientists across the globe. As the caretaker of Einstein's legacy on social media, I have the privilege of seeing this up close each and every day, whether I'm posting an article about how we have Einstein to thank for driverless car technology or a story about Loeb's search for extraterrestrials. "He has been an influence through-out my career," Loeb said.

A century after a historic eclipse allowed Einstein to use a telescope to prove his theories, another telescope—this one in a valley in the middle of nowhere in Green Bank, West Virginia—is continuing to stretch the imagination of what can be gleaned from the famed physicist's work. Free of cell phones and microwaves, just like in Einstein's day, the telescope is pointed toward the sky in search of something greater. What makes space glamorous, after all, is not the mere act of getting to where no one has gone before, but the imagination, ingenuity, and hubris required to do it. As astronomer Carl Sagan famously said, "In the deepest sense, the search for extraterrestrial intelligence is a search for ourselves."

As I got up to leave Loeb's office, I turned to ask him a question that's been on my mind since I first heard about his research. If he does spot an alien, what would he say to it? Loeb chuckled at the thought of a nebbish astronomy professor from Harvard stumbling upon E.T. "My young daughter asked me to bring the alien home if we ever find it," Loeb said. "My wife, on the other hand, said that if they ever offer me a ride on their spacecraft, I should make sure that I leave the car keys with her, and that they don't ruin the lawn in the backyard when they lift off."

WAR AND PEACE...AND EINSTEIN

"Politics is more difficult than physics."
—ALBERT EINSTEIN

t's not every day that celebrated actor Mandy Patinkin is wait-ing to get into my Zoom room. But there he was. And here I was, dressed for the occasion. Let's be honest, dressed may be an overstatement. I was still wearing pajamas from the waist down. But I had bothered to put on a cardigan and brush my hair.

I clicked Admit and a moment later, I was staring at Patinkin who, I'm happy to report, was also dressed from the waist up (in a black button-down shirt, if you're curious). His face was punctuated by his trademark salt-and-pepper beard, his bushy eyebrows adding to the expressiveness in his eyes. He was sit-ting in his cabin in upstate New York, the bookshelf behind him groaning under the weight of stacks of books, a record player, and countless albums. He blinked and looked at me, yet I was quiet for a moment, frozen not because of a wonky internet con-nection but because I, like so many people, have spent the past few years watching videos of Mandy Patinkin on social media, of him sitting in this exact room. Normally, I'm just a bystander, watching with amusement, stuffing ice cream in my face or half paying attention while I aimlessly re-order toilet paper on Amazon.

When the pandemic struck in March 2020, the actor and his wife of more than forty years, Kathryn, fled their Manhattan home and moved into this cabin they bought in the Hudson Valley. They found themselves living in the woods with little to do but wait out the apocalypse. Their son, Gideon, joined them and began recording fun videos of his parents squabbling, snacking, playing with their dog Becky, and doing mundane household chores. Watching Patinkin change the filter in his vacuum cleaner is nothing short of mesmerizing. At first, it was a way to pass time, but it soon become a healing tonic for a nation in pain. Two million fans across Twitter, Instagram, and TikTok tuned in to each video dispatch for some sense of normalcy and humor in a time of deep uncertainty.

So, there he was, and here I was, and the living, breathing Patinkin was apparently waiting for me to say something. Even Becky, who was napping on a chair behind him, perked up her head and looked at me quizzically. So, I finally flustered something.

"I can't believe I'm looking at the real Mandy Patinkin," I stupidly said out loud.

"I say the same thing to myself every morning in the mirror," he said, and capped it off with a hearty, guttural laugh.

Patinkin is exactly the same person in those social media videos as he was in my Zoom—relaxed, self-deprecating, said whatever is on his mind. We could've talked for hours about his award-winning performance in *Homeland* or his iconic role in *The Princess Bride*, a classic in our home growing up where we recited famous lines from the movie at every family gathering. But rather, you guessed it, we are here to talk about Albert Einstein. Allow me to explain.

Nazism was on the rise in the early 1930s and Albert Einstein, one of Germany's most famous Jews, was named an enemy of the state. He was barred from his teaching position at the Prussian Academy of Sciences in Berlin. When Einstein fled the country in December 1932, he told his wife, Elsa, "Take a very good look at it. You will never see it again." It was a premonition. The Hitler Youth tore apart his summer cottage in Caputh. The Nazis put a $5,000 bounty on Einstein's head. "I didn't know I was worth so much," Einstein joked.

By that point, Einstein denounced his citizenship in the country and fled to the United States. He was granted residency thanks to EB-1, a government program that gives priority to immigrants who have "extraordinary talent" or are "outstanding professors or researchers." It has since become known as the "Einstein visa" and occasionally shows up in the news because of foreign nationals who are seeking to take advantage of it. Melania Trump, a Slovenian model before becoming the First Lady, famously came to the United States on an Einstein visa.

Once in America, Einstein took it upon himself to use his celebrity and influence to help rescue his fellow German Jews.

He gave speeches and spoke at fundraising dinners. He even performed a violin concert for the United Jewish Appeal, a philanthropic organization. As Germany began its trek toward war, Einstein used his own money to help resettle Jews in Alaska and Mexico, among other places that were willing to take them. "I am privileged by fate to live here in Princeton," he wrote. "In this small university town, the chaotic voices of human strife barely penetrate. I am almost ashamed to be living in such peace while all the rest struggle and suffer." In the summer of 1933, Einstein formed the International Relief Association "to assist Germans suffering from the policies of the Hitler regime." First Lady Eleanor Roosevelt soon joined the cause. A similar group, the Emergency Rescue Committee, formed in France, and eventually Einstein merged the two together to become the International Rescue Committee. That group that Einstein founded still exists today, and the IRC is one of the world's largest refugee resettlement organizations. Its spokesperson is Mandy Patinkin.

It was 2015 when Patinkin was in Berlin shooting the fifth season of *Homeland*. At that precise moment in time, 125,000 Syrians were fleeing their country for a new life and a new beginning. "I'm looking at photos of these people in lines, trekking, and they reminded me of my ancestors, fleeing the pogroms of Poland and Belarus. We were all refugees," Patinkin said of his family that was lost in the Holocaust. "And I thought there but for the grace of God go I."

He wanted to go to a camp to visit the refugees, "not as a

celebrity, but just as a human being. I want to just go and walk with them and give them water and company and comfort and let them know I'm a human being who cares about them." So, Patinkin reached out to the International Rescue Committee. They invited him to Lesbos, Greece—the epicenter of where Syrian refugees were arriving—and the day after filming wrapped, he hopped on a plane. "The next thing I know, I'm there, and it changes my life. As Einstein said, 'A life lived for others is a life worthwhile.'"

Patinkin rushed to the edge of the sea and saw a boat coming to shore. "It's all overloaded and everybody's jumping off. And a person puts a little girl in my arms with a pink jacket," he recalled. She was limp, and Patinkin thought perhaps she was dead. "All of a sudden, I put my finger in her hand and I feel her squeeze my pinky." Through the chaos and confusion, he took her to a medical tent for help and eventually reunited the girl with her family. After that, Patinkin was hooked. He's since gone with the IRC to visit refugees in Uganda, Serbia, Jordan—as well as to the U.S.–Mexico border.

Patinkin's desire to help got turbocharged after the 2016 election of President Donald Trump, who slashed refugee admissions into the United States by 85 percent. "The damage done by the Trump administration was horrific," Patinkin told me. "Everybody in this fucking country is a refugee! And how do you have the nerve as the ancestor of anyone who made it here to say, 'Sorry, doors closed!'"

Patinkin told me a story about his Grandpa Max. "He used to say a Yiddish phrase, *dos redl dreyt zikh*, which means 'the wheel is always turning.'" Patinkin said. "If somebody on the bottom is knocking on your door and you don't open it up and welcome them and give them comfort, food and sanctuary and safety, when you're on the bottom, no one will open the door to you. It's just clear, moral, ethical behavior. For me, my job for the rest of my life, is to open the doors for others who are in need."

As the International Rescue Committee's executive vice president from 2000 to 2015, George Biddle traveled across the globe—to disaster areas and conflict zones—to support the group's humanitarian efforts wherever he could. "Historically, refugees were in camps, but now displaced persons are everywhere," Biddle told me. The nonprofit that Einstein launched was set up to help refugees from man-made problems, like war, but has extended its services to assist in the aftermath of natural disasters like earthquakes and tsunamis. "The IRC's mission is to go in and try and do the most effective lifesaving and life-preserving at the outset of an emergency, help to stabilize populations, and hopefully get them back to their homes and back on their feet so they regain control of their future." Biddle added: "We could provide clean water, we could provide primary health care, we could provide vaccinations. It could be small grants to just get people cash so they can get access to things they need."

Their efforts, and that of other similar non-governmental organizations, accelerated in the 1990s. "I have a theory about how this has all evolved," Biddle explained. "When the end of the Cold War occurred, there was a proliferation of sectarian and ethnic conflicts around the world, and there were many new humanitarian crises for the world to address. And I think, quickly, the United Nations and big governments recognized they needed help. And they increasingly turned to an NGO community, which was ready to jump in." Einstein's IRC was joined by groups like Save the Children, Oxfam, World Vision, and Doctors Without Borders. In 1990, the IRC's annual budget was in the neighborhood of $30 million; by 2000, it was over $100 million. And it continues to grow. "When I left the IRC in 2015," Biddle said, "the budget was just shy of $700 million. And there were probably 500 people at the headquarters in New York and 15,000 around the world." Its 2022 budget was well north of $1 billion.

The IRC is one of the largest resettlement agencies in the world. When a country like the United States agrees to take in refugees—say, Sudanese people who end up in Georgia or Arizona—the IRC is there to facilitate it all. "We greet them at the airport," Biddle said. "We arrange for a pre-rented home or apartment, and they have to have a warm cooked meal when they arrive. And the next day, they're immediately given a caseworker who is there to help them acclimate to their life in their new city, to help their kids get into school, to help them find jobs, get English-as-a-second-language classes, sign up for social benefits before they can get their job benefits going." The nonprofit even goes so far as to help them buy cars by providing loans. The IRC offers immigration services to help them move from refugee to permanent resident to, eventually, a naturalized American citizen. "You have to bring it down to the individual level," Biddle said. "You have to understand there's a human life in all that mass of humanity. Each one represents an individual whose life was completely upended through no fault of their own."

When Einstein established the IRC, dozens of theologians and thinkers and political leaders took his call to arms. "He was really the instigator who got these other people to then commit their time and energy," Biddle said. "He was the prescient one who could move people to act. They quickly realized, 'God, he's got a great idea. We've got to do something.'"

I asked Biddle what Einstein would think of what the IRC has evolved into today. "He would think it followed exactly what he thought it should do, which is do everything you can to help humanity on the edge—those who are most at a loss and most vulnerable. And do it as creatively, bravely, and in as disciplined a fashion as possible." Biddle said that when Bashar al-Assad, the president of Syria, attacked his country's civilian population in 2012, the IRC snuck food and critical medical supplies across the border. Each package was tagged with a unique QR code and was scanned on a smartphone by a local partner on the ground in Syria. Using GPS, an invention made possible courtesy of Einstein, the IRC was able to track where the package was moving. Since the technology was first developed, the IRC has opened its source code so other humanitarian aid agencies can use the system as well.

"Creativity means using all your brainpower," Biddle explained. It seemed very on-brand with Einstein.

Back in my Zoom room, Patinkin told me that he also took his cues from Einstein. "When Einstein arrived in America," Patinkin said, "he spent his own money to help people get visas, and then he joined the NAACP because he was in shock that everything he ran away from was happening right here with Black people being, as the Constitution originally said, three-fifths of a human being. It's just the same horror that the Jews were escaping." When the local Nassau Inn refused a room to Marian Anderson, an African American opera singer who was performing at Princeton University, Einstein opened the doors to his home. (A 2021 play *My Lord, What a Night* reenacts that event.) She returned to visit many times, and the two remained friends until Einstein's death.

"Being a Jew myself," Einstein said, "perhaps I can understand and empathize with how Black people feel as victims of discrimination," adding that, "My attitude is not derived from any intellectual theory but is based on my deepest antipathy to every kind of cruelty." Einstein once paid the tuition of a Black student. In 1946, he gave the commencement address at Lincoln University, the first school in America to grant college degrees to Black people. "The separation of the races is not a disease of the colored people, but a disease of the white people," Einstein said at the graduation. "I do not intend to be quiet about it." Patinkin shook his head in disbelief. "Einstein flees one situation, comes to safety, and immediately has the courage to speak out."

Einstein campaigned for civil rights and racial equality. In an essay called "The Negro Question," Einstein wrote, "The more I feel an American, the more this situation pains me. I can escape the feeling of complicity in it only by speaking out."

And speak out he did. He joined the effort to save the "Scottsboro Boys," nine Black teenagers from Alabama who were falsely accused of rape. In response to an uptick in racially motivated murders, Einstein went on a crusade to end lynching. Einstein worked with W. E. B. Du Bois, a founder of the National Association for the Advancement of Colored People (NAACP), to shed light on the plight of African Americans. Entire books have been written about Einstein's work to end racism. "Because he was so famous and situated in particular places at particular times, he managed to touch upon and influence a large number of different intellectual, political, and scientific movements," said Michael Gordin, who teaches a class called The Einstein Era at Princeton University.

Einstein served as the honorary president of the Children's Aid Society, a Jewish social welfare group in France. Four years after the war, children who had been relocated to the United States from a displaced persons camp in Europe knocked on the door to his home to thank him. Einstein was about to turn seventy, and he told the *New York Times* that it was a "magnificent birthday gift." A photographer on hand to capture the event, Philippe Halsman, was, according to *Time* magazine, "one of several people who the scientist himself had helped come to the United States in 1940 after the Nazis invaded France."

During one of Patinkin's many trips abroad, he met up with a family of Syrian refugees that had recently arrived in Germany. He asked the mother if she still had a lingering sense of fear. "She said, 'After what we've been through, we are afraid of nothing. I saw death behind me and life in front of me, and I just kept walking,'" Patinkin recalled. "It's the same thing with Einstein. After what he had been through, what he had just witnessed, he was not afraid to stand up for what's right—morally and ethically. He was intelligent in his soul. In my life, what he's known for is not the theory of relativity, but the theory of relatives."

It was a cold morning on February 24, 2022, when Russian President Vladimir Putin ordered an invasion of neighboring Ukraine. Air raid sirens could be heard throughout the country as martial law was enacted. Men of fighting age were required to take arms—while women, children, and senior citizens quickly packed up whatever belongings they could carry as they escaped the bombings. Within weeks, more than ten million Ukrainians fled, instantly creating Europe's largest refugee crisis since World War II. In Kyiv, the nation's capital, Anna Pysana, who was thirty-seven years old and single, had a decision to make.

In the days leading up to the invasion, Pysana couldn't sleep. "I had a feeling that something would happen," she told me in a video call, her bright orange hair punctuating the screen. For nearly a year, she had been having the same nightmare once or twice a week. "It was a dream about the end of the world." She described how she kept seeing the same scene replay over and over again. She was standing on the seashore and saw a hurricane approaching over the horizon. People on the beach were running away, but for some reason she stayed put as a flood encompassed the beach. She'd jolt awake each time. "I felt such a big fear, it was the worst feeling," she recalled. "I felt a tragedy inside."

Shortly before the war broke out, Pysana hopped on a train—not west out of the country to safer ground, but east toward Russia, to Sloviansk, her hometown. Her elderly parents had spent their entire lives there. They survived the invasion in 2014, when Sloviansk was the first city seized by Russian troops. This time, they hoped, would be no different. Besides, this is where all their friends were, they had a business and a farm and animals to tend to. At their age, they worried that they wouldn't be able to adapt somewhere new. So, they decided to hunker down and stay put. "It's quite complicated to explain why people don't want to leave," Pysana said. "They can't imagine a different life. So, they became a team with other people who stayed and they all support each other."

Pysana traveled back to her apartment in Odessa. But there it was no safer; bombs were going off all around her. "How long are we going to stay here and be in fear all the time?" she wondered. "It was crazy for my emotional health." She decided it was time to go, to leave her homeland. She grabbed whatever she could carry—a suitcase, a guitar, her camera equipment, and her yoga mat.

Pysana is Jewish on her mother's side, and, while she doesn't consider herself religious, she feels a kinship with others of her faith. After the fall of the Soviet Union, more than two million Jews left the region. The majority of them emigrated to Israel where, under the Law of Return, they were granted instant citizenship. Despite that mass migration, Ukraine still had one of the largest Jewish populations in the world—including thousands of Holocaust survivors. Pysana traveled to Israel many times, including once with a Swedish boyfriend, and had considered what life might be like there—but ultimately she decided it was not home. She attended school in Kharkiv, in northeast Ukraine, and put down roots there after graduation. She got a job at the Beit Dan Jewish Community Center where she was the director of its youth program. She coordinated events for teenagers and camps for families. She went to synagogue and attended Sabbath meals at the homes of local rabbis. "Judaism is a very important part of me," she said as she grasped a charm around her neck. It contained the Hebrew letters *chet* and *yud*, which spell the word for "life."

The community center where she worked is one of six in Ukraine funded by the American Jewish Joint Distribution Committee, commonly referred to simply as "the Joint," a New York–based nonprofit relief organization. A big part of its mission is to help resettle Jewish refugees. Pysana did not know it at the time, but her connection to the Joint would prove lifesaving.

In December 1930, Albert Einstein was in New York where he attended a meeting of the Joint's executive committee. "The Jewish people stand as a group who are always keenly responsive to the suffering in any portion of their people," Einstein said at the gathering, held at the Midtown Manhattan home of a philanthropist and attended by politicians, rabbis, and other community leaders. "Such suffering in any one portion of the group affects the entire body of Jews ... The Joint Distribution Committee has been instrumental in effecting considerable improvement in Jewish conditions of eastern Europe." Einstein added: "I sincerely trust that the Jews of the United States will continue this work, for with comparatively small sums of money, important reconstructive results can be obtained. I am confident that you will continue your vigorous interest. Only when the Jewish people retain their feeling of brotherliness, can we achieve self-respect and pride in our community." When Hitler took power, Einstein once again turned his attention to the organization. On June 15, 1941, Einstein thanked the Joint for orchestrating the rescue of one hundred children and giving them "a new lease on life." Decades later, Pysana was also a beneficiary of this kindness.

As the war in Ukraine began in February 2022, Jewish organizations from across the globe mobilized to help. Chief among them was the Joint, which already had a presence in the country. The group organized caravans of buses to drive thousands of Ukrainian Jews across the border. Pysana was among the passengers.

"I felt good and bad at the same time," she told me. She felt guilty that she was leaving her friends and family behind and was forced to flee because of the actions of someone else. "When I closed my apartment door for the last time, I had a feeling of saying goodbye forever, but also, I was thinking that I would eventually come back. But to what life would I come back to? It was unbearable."

Pysana is a photographer and filmmaker, and part of her wanted to stay behind to document the travesties of the war. But she knew she could be more useful elsewhere. Indeed, once the Joint helped her cross the border into Moldova, she immediately asked if she could volunteer at its refugee center to help other Ukrainians as they arrived. "I could offer support to these people, to give hope to them." And so that's what she did. For the next month, she stayed at the Joint's base, welcoming millions more of her fellow citizens as they swarmed across the border. Sometimes her work was simply handing them keys and showing them to their room or finding them clothes to wear. At other times, she was a de facto social worker, answering questions about paperwork and travel. "I met mostly old people," Pysana recalled. "There was this feeling like they didn't have a ground under their legs." It was, to say the least, an intense period. "I saw very different sides to human beings—from total disappointment to the very highs of deep soulfulness."

Recalling those prewar nightmares, of standing on the beach and being enveloped by a flood, Pysana called it a mystical experience. She said friends of hers in other Ukrainian cities told her they had the same dream. "It's our unconsciousness at work," she said. "Sometimes we don't pay attention to the things that can be important for a spiritual life. God is talking with us in different ways. It's just a mix of stuff that we can't explain."

At this point in the conversation, Pysana paused and looked at me, her green eyes piercing through the screen. "Einstein understood this since he was a very mystical person in general. People don't choose who we are. And some people like Einstein, they are a genius, and they are an instrument of God. Some people are like a clear piece of glass that God is looking through, and Einstein is a great example of this." Echoing the physicist, she said she feels "between worlds." When I spoke with Pysana, she was still unsure where life would take her. One of the first things she purchased when she crossed the border into Moldova was a book about surviving the apocalypse.

Lots of countries took in her friends—Poland, Germany, Sweden, just to name a few. She said it all boiled down to empathy. "If you were a poor person and then become rich, you understand what it is like to not have bread to eat or shoes to wear," Pysana said. "Maybe it's kind of a genetic memory." But, she added, "Empathy is something you have to work on. You have to educate people in this." She brought up World War II. "The German people were very educated, they were interested in classical philosophy, but intelligence does not always equal empathy."

She is inspired by Einstein, who was helping refugees until the day he died. "When Einstein started being involved with refugees, it was because of the Holocaust. Jewish people outside of Germany had more responsibility to help," Pysana said. "But I think even after the Holocaust, it's important to take that same feeling of empathy and apply it to other people and other groups that need it. You have to think that your life is very connected to the lives of those around you."

One of the nice things about being married to an academic is that I often get to tag along when Elizabeth is invited to speak overseas. Australia and France have been on the itinerary. She taught in Germany for a semester. A few summers ago, she published a research paper exploring how TV spoilers impact your viewing pleasure. She was asked to present her findings at a conference in Fukuoka, which I like to call the Cleveland of Japan. It's a perfectly nice place, beautiful in fact, but it doesn't draw the same crowds as Tokyo. What we did discover, though, was that it's not too far away from Hiroshima. And so, on a day off from the conference, we took a train to the first city ever hit by a nuclear weapon.

At 8:15 a.m. on August 6, 1945, a Boeing B-29 was flying above Hiroshima at 31,000 feet. The plane was named the *Enola Gay*, after the mother of its pilot, Colonel Paul Tibbets. It took forty-three seconds from the time Tibbets dropped the bomb until it struck the ground, pulverizing the city—instantly incinerating the population below, lighting everything on fire for miles. The city smoldered for days.

More than 70,000 people—30 percent of the city's population—died from the blast, and tens of thousands more in the month after from radiation. More than 90 percent of Hiroshima's doctors and nurses perished, most of whom were working downtown, where the greatest damage took place. Roughly two hundred Hiroshima survivors traveled south to seek refuge in Nagasaki. Three days later, on August 9, the United States dropped a nuclear bomb there as well.

As Elizabeth and I stepped out of the train, we couldn't miss one of the few remaining structures from that fateful morning. Thanks to its stone and steel earthquake-resistant design, the Prefectural Industrial Promotional Hall survived, sort of. What we were standing in front of was a hollowed-out hull of a building—windows missing, roof torn off—which gave visitors an eerie feeling. A monument to destruction. It was renamed the Hiroshima Peace Memorial and designated as a UNESCO World Heritage Site. It is in a permanent state of arrested ruin. A sign warns tourists that drones are banned. I guess the collective memory of something unknown flying overhead is still strong.

We walked a few blocks to the Hiroshima Peace Memorial Museum where we saw the impact that day had not only on the buildings, but also on the people who lived there. We learned of the charred bodies and smell of burned flesh that emanated from the city. One exhibit behind a glass box featured the detritus gathered from the wreckage—tattered clothes, a pocket watch, a child's bicycle. I've visited Holocaust museums before, which can strike a similar tone, displaying the relics of lives lost. As a Jew, I knew better than most what it was like to be the victim. But here, I felt different. I was an American. My people had caused this damage. Some lay the blame at one person in particular: Albert Einstein.

Much to Einstein's chagrin, many view him as the father of the atomic bomb. A 1946 cover of *Time* magazine showed an illustration of Einstein next to a mushroom cloud emanating from a nuclear bomb. The equation $E=mc^2$ is drawn in the smoke. His famous calculation gave future scientists building the bomb a tool to measure nuclear energy. What's more, as the Nazis rose to power, Einstein wrote a letter to President Roosevelt warning him that the Germans could be developing nuclear weapons and urging the United States to begin its own work on what would become called the Manhattan Project.

While all of that may be true, Einstein spent the last years of his life trying to set the record straight. "My participation in the production of the atomic bomb consisted of one single act: I signed a letter to President Roosevelt, in which I emphasized the necessity of conducting large-scale experimentation," he explained. "I was well aware of the dreadful danger which would threaten mankind were the experiments to prove successful. Yet I felt impelled to take the step because it seemed probable that the Germans might be working on the same problem with every prospect of success. I saw no alternative but to act as I did." It was confounding to Einstein: when he came up with the idea of the relationship between mass and energy in 1905, how was he supposed to predict the future? "There was never even the slightest indication of any potential technological application," Einstein said.

Alex Wellerstein, who runs a blog about nuclear weapons, thinks Einstein gets way too much credit. "The way I like to put it is this: $E=mc^2$ tells you about as much about an atomic bomb as Newton's laws do about ballistic missiles," he wrote. "At some very low level, the physics is crucial to making sense of the technology, but the technology does not just fall out of the physics in any straightforward way." He added: "Einstein did play a role in things, but that role wasn't as crucial, central, or direct as a lot of people imagine. If you could magically drop him out of history, I think very little in terms of atomic bombs would have been affected."

Bernard Feld was an MIT physics professor who admitted to being involved in the "original sin" of helping develop the atomic bomb. But, like Einstein, he later led an international movement among scientists to banish nuclear weapons. "The

use of the bomb against Japanese civilians was especially pain-
ful to Einstein," Feld wrote, "because he had formed very strong
friendships and a special fondness for the people during his visit
to Japan in 1922." Feld later became the president of the Albert
Einstein Peace Prize Foundation.

Einstein knew firsthand what nationalism and nativism could
lead to, calling it a "disease." He had seen what Hitler wrought in
Germany. "I am not only a pacifist, I am a militant pacifist," said
Einstein, who saw himself as a global citizen. "Nothing will end
war unless the people themselves refuse to go to war... We must
begin to inoculate our children against militarism by educating
them in the spirit of pacifism."

In May 1946, less than a year after the bombs struck Hiroshima
and Nagasaki, Einstein joined the Emergency Committee of
Atomic Scientists and became its chairman. The group, sadly,
did not see much success, but Einstein never gave up. One week
before his death—on April 11, 1955—Einstein launched a project with Bertrand Russell, a friend and philosopher, to reverse
the trend they were seeing of nations racing toward nuclear war.
Together they authored what's become known as the *Russell–
Einstein Manifesto* and called for scientists to assemble at an
annual conference to speak out against the development of weapons of mass destruction. They wrote that they don't represent
"members of this or that nation, continent or creed, but as human
beings," whose "continued existence is in doubt." They went on:
"We shall try to say no single word which should appeal to one
group rather than another. All, equally, are in peril, and if peril is
understood, there is hope that they may collectively avert it."

After Einstein's death, hoping to carry on Einstein's final wish, Russell and Joseph Rotblat, a Polish physicist, created the Pugwash Conferences on Science and World Affairs where scholars and public figures would gather regularly to work on ways to reduce the dangers of armed conflict. In 1995, fifty years after the bombing of Nagasaki and Hiroshima and forty years since the publication of the *Russell–Einstein Manifesto*, the organizers of the Pugwash Conferences won the Nobel Peace Prize.

I brought all this up with Patinkin, who sees Einstein's efforts for world peace still bearing fruit today. "One of Einstein's greatest gifts was his compassion for humanity," he said. "God brought gifts because of his intellect and genius that the world will benefit from through eternity." At the mention of a higher power, I inquired if Patinkin is religious.

"I'm spiritual," he told me. "I believe in Einstein's theory of relativity, which means that energy never dies. That's what Einstein proved. That energy in Abraham or Moses or my father or Steven Sondheim or Abe Lincoln or anyone on this Earth that you admire, the energy in their cellular matter is out there." He said his favorite word is *connect*. "If you want to connect to the six million lost in the Holocaust, all the Native Americans who've been lost, all the African Americans who've been lost, all those who've been treated inhumanely, my fellow human beings, you can reach them, you can gather them, you can be with them. They're all in the universe in some form."

But you can't have a conversation with them. So, I asked

Patinkin what he would talk about with Einstein when he meets him in the afterlife. "First I would say, let's go get some lox and bagels," Patinkin said with a roaring laugh. "We'd go sit down. And then I'd say to him, 'How do we address hate? How do we change a hateful heart? How do you undo hate? How do you change someone's fear of othering? How do we do that?' I can't figure it out other than acts of kindness."

Patinkin recalled the time he met Farhad Nouri, a ten-year-old Afghan refugee in Serbia. The two of them sat on a bench together, a film crew nearby, and Patinkin asked if there was a message the boy wanted to send to the world. "Yes," the boy said. "Refugees need kindness." Tears appeared in Patinkin's eyes as he retold me this story. "Einstein knew this. So, I would sit with Einstein, and I would say how do we teach a ten-year-old-boy's wisdom to a world that wants to embrace fear of others, xenophobia, hate, and employ violence toward others. How do we stop that? We live on a planet where, if we can protect it from being destroyed by human beings, chances are there could be enough space and resources, even as the population grows, to hold an embrace of that humanity long past our lifetime for generations and generations to come.

"And so, I would ask for his practical counsel. I don't want some intellectual idea, or some scientific explanation of the cosmos and physics that I can't get a hold of. Instead, I would ask: 'What do we do?' And I am certain that he would let me walk away from that lox and bagel sandwich with an action to take. I'm certain. He was a doer. A practical man."

As we wrap up the Zoom call, Patinkin explained why he

volunteered to be the face of Einstein's refugee organization and why he took the time to talk with me today instead of making a silly video with his dog for Instagram. It's because he wants to send a message to future generations. "If you have the privilege of having a platform for whatever reason, whatever field you're in, whether you're a writer or an actor, or a scientist, or a politician, or a businessperson, and you can reach one person, or many people—use that opportunity, use that privilege to speak out for what you believe in. And hopefully, it won't be about your own pocket and your own fame and your own need. Hopefully, it will be about things of desperate need—for survival of human beings and their beautiful families."

A smile came across his face. "It's about saying, 'Yes.' A guy who I never met named Benyamin, who's writing a book about Einstein? And there's a chapter about refugees? Say yes."

EINSTEIN'S MIRACLE YEAR

"Often I'm so engrossed with my work
that I forget to eat lunch."

—ALBERT EINSTEIN

P icture this: You're walking out of the grocery store, your arms full of bags, and the doors magically open. It's pouring rain outside, but you knew to bring an umbrella because your weather app warned you of the impending storm. After unpacking your groceries at home, your doctor calls with good news. The PET scan you had that checked for cancer came back negative. What do each of these technologies have in common? They were all made possible thanks to one person: Albert Einstein.

What's more astounding is that Einstein didn't come up with all the ideas that would lead to these inventions over the course of a lifetime. No, he came up with them all in one year, in 1905, when he was just twenty-six years old. At the time, he couldn't find a job teaching at a university. For a while, he tutored kids in algebra. But thanks to a connection from the father of a friend, he was able to gain employment as a patent clerk in Bern, Switzerland. That may not have been the academic gig Einstein was hoping for, but perusing all those patent proposals did hone his skills in one key area: he had an extraordinary ability to visualize physical situations based on technical material. The job at the patent office allowed him to be in this middle ground between theory and real-word applications. "His visual imagination allowed him to make conceptual leaps that eluded more traditional thinkers," wrote Einstein biographer Walter Isaacson.

Nobody paid Einstein much attention, if any at all. But that would soon change, and at a remarkable clip. He would become so prolific over the next several months that historians simply call it Einstein's "Miracle Year." (Or if you prefer the Latin, Einstein's *annus mirabilis*.) It was something the scientific community hadn't witnessed since 1666 with Isaac Newton. During 1905, Einstein authored four revolutionary research papers that would turn scientific orthodoxy on its head. And more than a century later, his ideas are still bearing fruit in modern-day inventions we all come in contact with on a daily basis.

"He was an ambassador whose theories laid the groundwork for the century's great achievements: electronics and nuclear power, space travel and television," explained Carolyn Abraham, who wrote a book about Einstein's brain. "His work underpins our digital age. All these things that he dreamed up with nothing more than his head and a piece of paper and a pencil has really propelled our civilization in profound ways that continue to be meaningful and relevant and game-changing."

If a scientist had written just one of these papers, it would have been a career-defining moment, one which they could rest their laurels on for decades. But Einstein somehow managed to write four such papers in quick succession —and he did it all in his spare time, at night in his second-floor apartment, while his wife and one-year-old son, Hans Albert, slept in the other room. He feverishly wrote by the light of an oil lamp. Einstein's miracle year was so unique, it has its own Wikipedia page. The Einstein estate sells T-shirts and hoodies with "1905" emblazoned across them.

I kept reading about Einstein's miracle year and how it upended science as we know it. But I, for one, didn't know enough about physics to grasp its true significance. I tried reading about it—in a book called *Einstein for Dummies* no less—but my eyes glazed over as I stared at what was literally Greek letters and numbers. After wrestling with the mass-energy equivalence, I finally solved it by figuring out that I could throw the book away. What I needed was a guide, somebody to help explain these four papers to me, your Average Joe, the likable idiot. I failed science. Let me rephrase that: I despised science so much that, in college, I just stopped attending classes. Einstein dropped out of high school and one of his college professors called him a "lazy dog," so I guess I'm in good company. I needed someone who had experience explaining Einstein to the masses. Fortunately, I had the perfect candidate right in my small town.

It was late afternoon on a beautiful spring day here in Morgantown, with clear blue skies and near-perfect weather. I was taking a walk with Sean McWilliams down his block to pick up his young daughter from the bus stop. McWilliams, dressed in a blue polo shirt and khakis, is a physics professor at West Virginia University, with a particular obsession for Albert Einstein. He teaches an entire course on general relativity (he graciously allowed me to sit in on a few lectures) and organized a "Celebrating Einstein" conference on campus, which I had attended. It featured, among other things, an Einstein lecture performed through interpretive dance. We broadcasted it live on Einstein's Instagram page. Perhaps more impressive, though, was that McWilliams made international headlines for being part of a team of scientists who were finally able to confirm the last prediction of Einstein's general theory of relativity. It was something even Einstein himself couldn't accomplish.

To understand what McWilliams and his colleagues discovered takes a little bit of a science lesson: When two black holes merge, they create a ripple in space known as gravitational waves. The laws of physics explain that there are only two kinds of waves that can travel across the universe and bring us information about what's far away: electromagnetic waves and gravitational waves. It was Galileo, more than four hundred years ago, who built the first optical telescope and introduced the world to electromagnetic waves. And it was Einstein who described gravitational waves in 1916. It was a concept he came up with that he could visualize in his mind, but he wasn't able to detect these waves in real life. For a century, people tried, but to no avail. Proving their existence sparked generations of researchers.

Eventually, technology caught up with those ambitions. In 1994, three scientists—Barry Barish, Kip Thorne, and Rainer Weiss—devised a plan based on an idea they had been conceptually discussing since 1968. They built an L-shaped tunnel in the middle of an open field; it was massive—2.5 miles long in each direction—and took three years to construct. They built two of these contraptions, dubbed LIGO for Laser Interferometer Gravitational Wave Observatory, one in Washington state and the other in Louisiana. Inside were lasers and mirrors and other contraptions that could detect these ripples in space-time. The machine is so microscopically calibrated that it is capable of sensing a fluctuation 1,000 times smaller than an atomic nucleus. Like I said, I'm no scientist, but that seems pretty small. "The fact that humanity could build a ruler that accurate astonished me," McWilliams said of LIGO. When the trio asked the National Science Foundation for $1 billion in funding, "everybody thought we were out of our minds," Weiss recalled. But they got the money, and it would soon become one of the largest projects ever funded by the NSF.

Shortly after they flipped the switch and turned LIGO on, something remarkable happened. On September 14, 2015, two black holes, each thirty times more massive than the sun, collided in space. The collision, as Einstein had predicted, emitted ripples of energy in the form of gravitational waves, which traveled across the universe. The waves released a rising chirping sound, a distinct audio signature of this particular celestial event. Signals as strong as this one had arrived regularly on Earth for billions of years. But, for the first time, humanity had the necessary technology to detect these siren songs.

Back here on Earth, most of us were plausibly distracted: the guitarist for REO Speedwagon had just died, Serena Williams lost the U.S. Open, and Donald Trump was getting a rock-star greeting at a presidential campaign stop in Iowa. But the two sensors inside the LIGO tunnels startled awake at four o'clock that morning to the vibrations of those ripples. The perfectly positioned mirrors shook ever-so-slightly. "Data was streaming, and then 'Bam!'" recalled one of the scientists who clocked the change. As the *New York Times* later described it: "That simple chirp, which rose to the note of middle C before abruptly stopping, seems destined to take its place among the great sound bites of science, ranking with Alexander Graham Bell's 'Mr. Watson—come here' and *Sputnik*'s first beeps from orbit." Dr. Thorne called it "the most powerful thing that humans have ever observed." The discovery gave us an entirely new way to see the universe. It was the dawn of a new age in astronomy.

A few months later, on Christmas Day, LIGO detected a second pair of merging black holes. The three scientists sent all the data they had collected from both collisions to one thousand of their colleagues across the globe—including McWilliams here in West Virginia—to verify that what they had heard were in fact gravitational waves. They all agreed that it confirmed Einstein's theory and completed his vision of a universe that was not stagnant, but one that could stretch, contract, and jiggle. The scientists published a paper and announced their findings in 2016, exactly one hundred years after Einstein first predicted the existence of gravitational waves. And then in 2017, like Albert Einstein himself, they were awarded the Nobel Prize in Physics.

"People from all walks of life understood the significance," said McWilliams. "We were made to feel small, just like I was, by the thought of these unimaginable objects racing at unimaginable speeds, of this cataclysmic collision billions of years in the past, and of its faint echo, racing toward us across an endless expanse of time and space."

In the ensuing years, McWilliams fine-tuned the mathematical calculations involved in listening for gravitational waves. The scientists have since spotted even more black hole collisions. Now, NASA has enlisted McWilliams's help to build a gravitational wave detector in outer space. "If we can put a detector in space, away from the ground noise, then we can discover so much more."

Which is all to say that if someone could explain the four groundbreaking research papers of Einstein's miracle year to me, it would be this guy. He was a physicist who had literally finished what Einstein had started. Sure, McWilliams was opening up the cosmos for all to see, but he found time in his busy schedule to give me an introductory lesson on Einstein's miracle year. As we walked down his street, past sugar maple trees and the homes of other professors, I reminded McWilliams that I needed him to be gentle: teach it to me like I'm a kid at his daughter's elementary school. I think I may have used the phrase "in dummy terms, please." And so, he did.

Albert Einstein grew up around people who thought about light for a living: his father and uncle ran a company that installed electric streetlights in Germany. So, perhaps it shouldn't come as a surprise that the first paper Einstein wrote in 1905 described the photoelectric effect—and it laid the groundwork for so many technologies that we use every day. Before Einstein came along, people had thought that light was just one continuous wave. They assumed if you shined a light onto something, the surface would absorb the light. But Einstein said light was actually comprised of a bunch of discrete packets, what McWilliams called "little ball-like objects." If you were to shine a bright light at a surface, you could cause these packets to move. They could, for example, now have enough energy to dance down a wire and cause an electrical current. The best example of this phenomenon would be a remote control. When you shine the light of the remote control at your TV screen, the light particles hit the TV, get converted into energy, and can change the channel. So, the next time you're perusing Netflix's vast library of content, deciding between *Stranger Things* and the final season of *Grace & Frankie*, clicking the arrows on your remote control while you're nestled on the couch, remember that Albert Einstein is the one that allows you to be so lazy.

The photoelectric effect can be seen in so many modern-day technologies beyond just remote controls. Fiber-optic cables that snake their way around your city into people's homes, which allow us to make phone calls and surf the internet, convert light particles into electricity. In that sense, you could say that the paper Einstein wrote in 1905 helps power your Google searches, your dating profile, and every YouTube video you've ever watched. In 2021, Japanese scientists announced they had used this theory of converting light into energy to create an entirely new treatment for cancer. "Every photoelectric cell can be considered one of his intellectual grandchildren," noted one Einstein scholar.

Let's go back to our grocery store scenario from the beginning of this chapter. We're all familiar with the automatic doors that open as we enter the store. The way most of those work is that there is a device on top of the door that is beaming down a ray of infrared light, invisible to the naked eye. When nobody is in front of the door, the light simply hits the ground and bounces back up to the top of the door. But when you walk in front of the door, the beam of light hits you and the little packets inside the light get excited, start dancing around, and drive an electrical current that tells the door to open. Light, as Einstein described in 1905, can be turned into energy. And that energy, in turn, can drive an electrical current. Indeed, any technology that uses a motion detector—like a burglar alarm or automated lighting in a smart home or elevator doors knowing not to slam together when someone's hand is in the way—is employing the photoelectric effect. So many electronic devices—DVD players and police scanners and solar panels, just to name a few—operate on this principle of converting a ray of light into an electrical signal. It's also what allows digital cameras to convert photos into electronic files, as an image sensor converts light into discrete computer code. Indeed, Einstein patented an auto-exposure camera five years before Kodak.

Let's stay inside the grocery store for a moment. McWilliams explained to me that any device that operates on bouncing light off something finds its roots in Einstein's photoelectric effect. Take, for example, the scanner at the checkout lane. There is a solid infrared light there at all times. But when I move a box of Fruity Pebbles over the scanner, the light reflects off the box, the various particles in the light dance around and send an electrical message to the machine (which should be: This middle-aged man doesn't care about his health and is buying kids' cereal).

A supermarket in Ohio, in June 1964, became the inaugural store to use such a scanner. The first item to pass over that glass plate? A pack of Wrigley's chewing gum. "And that gum," wrote food journalist Corey Mintz, "with the aid of a Spectra-Physics model price scanner that now lives in the Smithsonian, became the Neil Armstrong of groceries, the first item to be scanned and sold, transforming the analog systems of inventory control until then managed by paper, pencils, and human memory." Lest you think every grocery store invention comes courtesy of Einstein, allow me to quickly disavow you of that notion. Sylvan Goldman came up with the idea of the shopping cart in 1937 for his Oklahoma City chain of stores, Humpty Dumpty. And many people weren't that excited about it; at first, it was rejected by men as too effeminate. Some said it was like pushing a baby carriage. To convince people otherwise, Goldman hired male actors to walk around the store and pretend to shop.

Inventing the technology that powers remote controls and radios and televisions and burglar alarms and motion sensors and that thing that scans your luggage at the airport—all of that would've been enough. But there is more: Think about lasers. They are basically the photoelectric effect on steroids. What we've been talking about so far are activities that happen at close range—you walk in front of the door at the grocery store, you scan your cereal box at the checkout counter—but lasers take that process and repeat it practically indefinitely. The light from a bulb diminishes the further away you walk from a lamp. But a laser will retain its strength. "The beam doesn't get weaker with distance," McWilliams said, as we waited at the bus stop. Einstein himself postulated the concept of a laser in a later paper in 1917.

Lasers are used in everything from eye surgery to military weaponry to the space program. As we saw in a previous chapter, Dr. Loeb is using lasers to push tiny spaceships outside our solar system in the search for aliens.

A relatively new technology I referenced earlier known as lidar (light detection and ranging) helps scan 3D objects in your environment. Lasers are sent out, bounce off a surface in the environment, and come back to the receiver, allowing it to effectively ascertain how far away an object is. Lidar is now used to help driverless cars figure out their surroundings and prevents them from hitting other cars or slamming into a building. Inside your home, lidar prevents your Roomba from bumping into the couch. "The photoelectric effect lays the foundation for these lasers," McWilliams said.

Einstein may be most famous for $E=mc^2$ and the theory of relativity, but it was the photoelectric effect that eventually earned him the Nobel Prize in 1921.

A month after Einstein printed his first paper, he was already publishing his second. This one was about a concept called Brownian motion. McWilliams said the easiest way to understand this is by looking at someone who's had one too many drinks. "If you have a drunk person standing at a lamppost, they can stagger left or right," he explained, as we ourselves waited at a lamppost for his daughter's bus to arrive. "And there's usually equal probability that they'll stagger in either direction. You can describe statistically where they're likely to be. And most likely, the average drunk person is back at the lamppost because they're equally likely to make a left step or a right step."

Sure, given enough time, and enough beers, the drunk guy could stagger away from the lamppost. Maybe he'll saunter over to his ex-partner's house and beg them to take him back. That's certainly possible. You could glance over at the lamppost and the drunk guy might be gone. But if the drunk guy takes one hundred steps, on average it will be fifty steps to the left and fifty steps to the right, and he'll likely still be in the vicinity of the lamppost. Einstein perfected the mathematical equation underlying that concept.

Weather prediction, for example, employs the concept of Brownian motion. "At least for short periods, you can model a lot of weather as a Brownian process or as a random walk," McWilliams explained. So, you could look at today's weather and be pretty accurate about predicting tomorrow's. But after time, those initially small random excursions become greater, which is why you can't forecast the weather for six months from now. Just like you couldn't go back to that lamppost in six months and still expect to see the drunk guy standing there.

The mathematics of Brownian motion can also be applied to the stock market. Like the example of the drunk person, stocks move slightly around—they go up, they go down. It's a random walk around certain trends: such as supply and demand, greed and fear, news and the state of the world. When a stock is volatile, it means it's taking a big excursion away from the mean. (Like the drunk guy leaving the lamppost to see an old flame.) "You could lose an awful lot of money," McWilliams said. But if the random spikes are smaller, as described in Brownian motion, then the worst-case scenario is less. "And so maybe you would want to invest in a more conservative portfolio."

I was particularly surprised when McWilliams told me that most of the products in my bathroom come courtesy of Brownian motion. Pharmaceuticals, for instance, have an active ingredient and several inactive ones. "You can't really mix them at the atomic level. You can't take one atom of this thing and three atoms of this other thing and six atoms of something else." But thanks to the concepts explained in Brownian motion, you can have an active ingredient that is buffeted on all sides by inactive ingredients. Imagine energy batting around an atom. That atom is equally likely to be kicked left, right, front, back, up, and down. On average, the suspension remains exactly how the drugmakers intended it. "You can just let the material mix itself on an atomic scale," McWilliams said. "The fact that Brownian motion happens means everything gets nice and uniformly mixed." Without that, you wouldn't be able to control the rate of release, or you might take a drug and get six times the dose you were expecting. The same thing would apply to other chemical mixtures in your bathroom, like toothpaste and shaving cream. The commercial production methods of these household items all use diffusion processes, first explained in Einstein's 1905 paper on Brownian motion.

Weather predictions, stock market modeling, producing great shampoo—these are all important innovations. But perhaps the most integral aspect of Brownian motion, this discussion of molecules bouncing this way and that, is that it confirmed the very existence of atoms. Although atoms can't be seen with our own eyes, their effects on any given particle—the perfectly mixed shampoo product—could be viewed. Not too shabby, and yet Einstein did not see himself as a revolutionary. "Einstein took every opportunity to disavow it," wrote Gerald Holton, a science historian. "He saw himself essentially as a continuist," someone who was merely building upon the theories of scientists who came before him.

Einstein's third paper of 1905 was the introduction of the special theory of relativity (not to be confused with the general theory of relativity, which was published eleven years later). Before Einstein, Newton imagined that there was a universal timekeeper in space. If you measure a second here on Earth, and measure a second in a distant galaxy, it would all be the same. "If I was carrying a ruler around," McWilliams explained, "the length of that ruler would be the same no matter where I am." The reason for this is Newton thought that light traveled instantaneously. So, if you're looking up at the stars in the sky above, then you'd guess that's what the star looks like right now.

But that notion simply wasn't true. Jupiter, for instance, is not always at the same distance from Earth. If you look through a telescope at Jupiter when it is at its furthest point, the light is taking a little bit longer to reach us. And so, we're seeing Jupiter not where it actually is, but where it was some finite amount of time ago. Up until Einstein, there was no experiment that said the speed of light depended on anything. "It was the same no matter what direction you were moving, how you were moving," McWilliams said. "Everybody measured the same speed of light." Except, as Einstein pointed out, it's all relative.

Imagine a man standing in the aisle of a moving bus. He's juggling balls in the air. The man is standing still and so juggling the balls is not that difficult (assuming he's a half-decent juggler). Now imagine you're on the sidewalk watching this. The bus zooms by and you catch a glimpse of all these balls in the air. People's perceptions of speed and direction would differ depending on where they were (inside the bus or outside of it). Einstein was able to figure that out. "The math is no more complicated than algebra," McWilliams said. "Your typical high school sophomore could pretty easily handle it." I didn't have the heart to tell him that not every student was as smart as that. My mother was my high school algebra teacher, and I still performed horribly in the class. Dr. Hanoch Gutfreund, the director of the Albert Einstein Archives, made me feel a little better when he wrote of the special theory of relativity that "it is a puzzle in itself that so many idolized Einstein for an achievement of which they understood almost nothing."

Decades after Einstein's miracle year, in the early 1960s, a team of computer scientists in the U.S. Navy picked up on this idea. They were measuring the relative distance between Pluto and Neptune and using the calculations described by Einstein in his special theory of relativity. Out of that research, the GPS navigation system was ultimately invented (which I explained in greater detail in an earlier chapter). Combining special relativity with quantum mechanics led future scientists to discover positrons, which laid the groundwork for modern medical innovations like the PET scan, allowing doctors to detect cancerous tumors or illnesses likes Alzheimer's and dementia.

At this point in the tutorial, McWilliams's daughter hops off the bus and joins us for the discussion of Einstein's final theory of his miracle year—a three-page document published on November 21, 1905—which described the relationship between mass and energy. Indeed, what Einstein realized was that mass and energy were one and the same thing. He came up with an equation to explain this phenomenon that is so surprising that he wondered if the universe was playing a practical joke on him. As he wrote to a friend: "The contemplation is amusing and attractive, but I don't know if the good Lord is laughing at it and leading me around by the nose." Einstein concluded that energy equals mass times the speed of light squared, but you may know it as the world's most famous equation: $E=mc^2$.

What Einstein said was that any object—from a pencil to a book to a bomb—has energy locked inside of it. For example, take a flashlight and place it atop a highly sensitive scale. If you turn the flashlight on, the reading on the scale will go down and down. Because the flashlight is emitting light, and that light carries away energy. And if it carries away energy, it carries away some of the mass of the flashlight, and therefore the scale's reading will go down.

Furthermore, let's look at two objects, each with the same amount of energy stored inside. For our purposes, let's give that energy a value: 1. If you smash these objects together, you would think that you would now have a new object that emits 2 bits of energy. However, what Einstein said was that the act of the collision itself also contains energy. So that when you smash an object with 1 energy together with another object with 1 energy, you actually have a lot more energy than you realize. (His mathematical equation explains just how much. The resulting number has lots of zeros after it.) So, the amount of energy an object could contain is not really based on the size of that object, but by how many smaller masses were slammed together to make that big object. In theory, you could continue to fire tiny little objects together until you build up an object that has enough energy locked inside that it could blow up an entire city. Hence, the equation of $E=mc^2$ allowed future scientists to build nuclear weapons and atomic bombs (a topic I wrote about in the previous chapter). Einstein would later auction off a copy of one of his 1905 papers to help fund a war bond drive in America. It sold for $6.5 million and was eventually donated to the Library of Congress.

Einstein's massive output during his miracle year may seem superhuman. To be honest, I'd be happy to publish just one thing each decade that people remember. Yet study after rigorous study has found that the concept in sports known as having "the hot hand" can occur across industries. Hollywood film director Rob Reiner had a string of back-to-back hits—*This Is Spinal Tap* (1984), *The Sure Thing* (1985), and *Stand by Me* (1986)—that allowed him to convince wary studio executives to let him make his next movie, a unique comedy-romance-fantasy mashup: *The Princess Bride* (1987). It became an instant classic. When a plague ravaged Europe in the early seventeenth century, it offered William Shakespeare the perfect opportunity to stay inside and write. Think of it as his very own pandemic project. It was during that time that he penned *Romeo and Juliet, King Lear, Antony and Cleopatra,* and *Macbeth*—a hot streak by any stretch of the imagination. Between 1914 to 1918, while World War I was being fought across Europe, Einstein had his most productive years ever, publishing fifty-nine scientific papers.

Ben Cohen, a sports reporter for the *Wall Street Journal* who happens to share my name, wrote a book on this topic called *The Hot Hand: The Mystery and Science of Streaks.* It's a meditation on the magic touch. "There is something to this idea that success begets success," he told me. When he's personally gotten into a groove with his own writing, he said he has to remind himself "that this is really the time to work a little bit harder, and maybe stay at the office later, or to wake up earlier, to go to sleep later and just churn out as much as possible because the one thing that we know about the hot hand is that it runs out. It does not last forever."

For Einstein, however, that notion would prove untrue. He

actually published a fifth paper during his miracle year. Eleven days before submitting his work on Brownian motion, he turned in his doctoral dissertation. It was a mathematical equation that deduced the size of molecules that has modern-day applications in everything from cement mixing to deodorant. The overproductive Einstein later published a paper explaining why the sky is blue. *We get it, Albert, you're a high achiever.*

"It's extraordinary," McWilliams told me as we arrived back at his house. "The vast majority of physics has been incremental. If I tweak this or turn this knob, then I maybe can explain a very slightly different phenomenon. But Einstein was able to do so many different things that are so completely out of the box. It's not surprising that he's had such a lasting impact—because his ideas weren't incremental. All of them were dramatic shifts from the existing paradigm."

BRAND EINSTEIN

"The cult of individuals is always,
in my view, unjustified."

—ALBERT EINSTEIN,
AFTER VISITING AMERICA FOR THE FIRST TIME

John Reznikoff holds the Guinness World Record for the largest collection of celebrity hair.

The artifact dealer has sold a small clump of Elvis's hair for $15,000. Twelve single strands of Michael Jackson's hair sold for $2,000. A barber in Ohio, who had given a buzz to Neil Armstrong, secretly pilfered some of the astronaut's hair and sold it to Reznikoff for $3,000. (Armstrong later threatened to sue.) Reznikoff has human hair samples from Beethoven, Napoleon, Marilyn Monroe, and Ava Braun. He's apparently got Geronimo's entire ponytail. And, yes, he's got some strands of the iconic shock of white hair that once belonged to Albert Einstein.

"It's a very Victorian tradition kind of thing," Reznikoff told me when I ask him how the heck is there a market for famous folks' follicles? "One hundred years ago, when somebody famous came to town, you didn't necessarily ask for their autograph. You asked for a lock of their hair. But also, when children were born and people died, it was also traditional to save a lock of hair." I admit to him that when my dog, Tivo, passed away I saved some of his hair in an envelope. It's now in a safety deposit box at the bank.

Reznikoff also pointed out the literal staying power of a strand of hair. In 2018, archaeologists in Peru discovered the nine-thousand-year-old skeleton of a young woman. "What's left?" he asked matter-of-factly. "The bones and the hair."

These famous hair samples are par for the course for Reznikoff. As founder of the relic-collecting University Archives he literally trades in the macabre. Tall and thin, Reznikoff cuts a lanky figure. His eyes, always at a caffeinated bulge, belie a child's sense of wonder—or of the Crypt Keeper himself.

He began collecting when he was eight years old. His grandmother, a Holocaust survivor, handed him an envelope of stamps and he's been hooked ever since. "I still have those stamps in my vault."

In *The Voyage of the Damned*, the movie about the sailing of the SS *St. Louis* filled with German Jewish refugees, Faye Dunaway portrayed Reznikoff's great-aunt. "That history is so important to me and ties together with what I do today," he said. One of the earliest artifacts he was asked to appraise was a collection of love letters and poems written by Nazi propagandist Josef Goebbels to his mistress, who happened to be Jewish.

It may seem odd that Reznikoff, himself Jewish, would make money on Third Reich relics. "You can't look at something done by Hitler and not have a range of emotions," he explained. "I believe the material should be preserved, but it is repugnant." He liked to quote the famous philosopher George Santayana: "Those who cannot learn from history are doomed to repeat it."

It was a noble and high-minded catalyst for Reznikoff, perhaps a real-life Indiana Jones, to get into this bizarre line of work, but it was something more tangible that kept him in— money. For decades, dynastic families like the Forbes clan had been famous collectors of historical artifacts. They sparked a national frenzy—and forever changed the marketplace for high-end collectors—when, in 1985, they purchased a two-hundred-year-old Château Lafite Bordeaux that purportedly belonged to Thomas Jefferson for a record-breaking $156,000. It was, simply put, the world's most expensive bottle of wine. (It was later discovered to be a fraud.)

The Forbes family bought the stovepipe hat and opera glasses from the night of Lincoln's assassination. They purchased Imperial Fabergé Easter eggs from Czarist Russia, a collection second in size only to the Kremlin. They set up a museum of these artifacts in the lobby of their Manhattan offices for all to see. Arthur Sackler, the pharmaceutical industry billionaire, collected so much Chinese art that he had to buy a second house just to store it.

Historical artifacts—Civil War treaties, ancient manuscripts, old paintings—had been the currency of wealthy collectors as far back as history would allow. The tradition, like the items themselves, had stood the test of time. But Reznikoff saw something more: he saw an opportunity.

The Wall Street greed of the 1980s and the dot-com boom

during the 1990s saw a rise in the nouveau riche—those for whom wealth didn't come from a bloodline and the luck of a last name like Rothschild or Vanderbilt, but instead from a modern-day gold rush. With that, came an insatiable appetite, a desire for one-upmanship. No longer would a Renoir suffice. These new collectors were looking for something more, something sexy. You know, like Albert Einstein's hair.

Enter John Reznikoff. He understood the power of hype and the allure of the weird. His fascination with celebrity hair garnered him headlines, and plenty of orders. Lots of people wanted in. But there's an obvious dilemma: the law of supply and demand dictated that he needed more. I mean, he couldn't just call a warehouse and order more of Einstein's hair. Reznikoff simply didn't have enough hair to go around.

So, he did the next best thing. He mass produced it.

He sent the celebrity hair he had on his shelf to a laboratory. There, scientists took the lock and used pressure and heat to produce carbon; the carbon was then heated until it became graphite; and then that material was pressed to make diamonds from the celebrity DNA. He was an alchemist of fame. People could now own a piece of Michael Jackson's hair. Bill Panagopulos, a competitor of Reznikoff's in the memorabilia market, joked: "There's enough George Washington hair to carpet your living room."

The new company was christened LifeGem. "It would almost be remiss of us not to make a diamond and offer that to the world," said Greg Herro, Reznikoff's friend and a co-founder in LifeGem. Diamonds were just the beginning. Various jewelry at all price points were soon available.

"Do I have a *Jurassic Park* mentality and think that I'm going to have a dinner party with all the greatest people in the world? No, not in my lifetime," Reznikoff admitted. "But I do think we're going to be able to determine what people's genetic backgrounds were, their history, and where they were from. We're going to be able to '23andMe' everyone pretty soon—including dead people."

Walk into University Archives and don't expect a wood-paneled, globe-filled library, with books on shelves so high you need a ladder to reach them. It's not the Explorers Club meets the National Geographic Society. It is very apparent this is exactly the opposite. It's located in what can only be described as the most nondescript office park in Wilton, Connecticut. When I walked in, I half expect Steve Carrell's character from *The Office* to step out to greet me.

Small black chairs on wheels are parked in front of bland brown desks. Reznikoff gave me a tour, opening up the drawers of vast metal filing cabinets. It's like he was a doctor grabbing a patient file, but instead the manilla folder contained a sheet of paper encased in a clear plastic cover. It was an autographed letter signed by Einstein.

"The interesting thing about Einstein, I think, as a market, is it is not shallow," he said, adding that no matter the amount of Einstein memorabilia, there will always be people interested in buying it. "I'd say the market for Einstein goes up 20 to 30 percent a year."

In a 1987 auction, a handwritten paper where Einstein scribbled out his famous formula, $E=mc^2$, sold for $1.2 million. At the time, the price was a record auction for any manuscript sold in the United States and for an unillustrated text manuscript sold anywhere in the world.

Einstein memorabilia continues to flood the marketplace, most of it bought and sold at auction. In 1945, Einstein got an X-ray of his skull; in 2010, it fetched $38,750. In 2015, a batch of Einstein's letters yielded $420,000. In 2016, a letter that Einstein wrote to his son was auctioned for around $100,000. An autographed copy of the classic photo of Einstein sticking out his tongue sold for $125,000. A violin once owned by Einstein sold for $516,500. A 2017 auction of eight letters written by Einstein netted $210,000. At another auction that same year, Michael Jackson's best friend—mentalist Uri Geller—purchased a letter Einstein wrote for nearly a quarter million dollars.

Einstein once sat for a sculptor who made a casting of his head. That wax mold, later autographed by Einstein, was sold for more than $20,000 and ended up at the Ripley's Believe It or Not warehouse. An autographed copy of Einstein's doctoral dissertation sold for $19,000. The genius's iconic pipe sold at a Christie's auction in 2017 for more than $70,000.

When Albert Einstein was staying at a Japanese hotel in 1922, he found himself without any cash for tips. So, he scribbled two notes and handed them to the bellhop, reportedly telling him, "One day these will be worth something." In 2017, Einstein's prediction came true: those two notes sold for $1.56 million, the largest sale of Einstein memorabilia up to that point in time.

That record didn't last for long. In 2021, a rare fifty-four-page manuscript in which Einstein had written his thoughts about the general theory of relativity sold at a Paris auction house for $11.4 million.

"Why are so many people obsessed with Einstein?" I asked.

"One word: smart," Rezinokff replied. "He is the iconic poster child for a person who has intelligence. People would say, 'Well, you're no Einstein,' if somebody is not smart. As simply and as often as somebody would say, 'Well, put your John Hancock here.' There's a reason for that. John Hancock wrote his name enormous on the Declaration of Independence. So, the concept of a John Hancock as a signature is the same as Albert Einstein for intelligence."

What's more, Reznikoff explained, is Einstein's mass appeal. "People who are interested in science, people who are interested in autographs, people outside of my hobby who just admire him. He draws from everywhere."

While Reznikoff has bought and sold plenty of documents over the years, his bread and butter are three-dimensional relics. "I know this sounds crazy," he said, "but when I hold them, I feel like I almost channel that person." His cluttered office—littered with the historic detritus of lives others have lived—is a madcap collection that would make the world's most bizarre hoarder envious: He has Ernest Hemingway's typewriter and Annie Oakley's gun. Shoeless Joe Jackson's baseball bat. Meyer Lansky's betting slips. Reznikoff sold a childhood toy of Einstein's for $27,500. In total, the self-described treasure hunter has bought and sold more than $200 million in collectibles.

Presidential iconography is of particular interest to Reznikoff, who studied political science and anthropology at Fordham University. He claims to have a tuft of Abraham Lincoln's bloodied hair from the night he was shot. His hair collection includes samples from Eisenhower, Nixon, and Reagan. When a man stepped forward claiming to be the illegitimate son of JFK, Reznikoff offered investigators a locket of JFK's hair so they could use DNA to debunk the claim.

The celebrity hair collector also peddles random items like spark plugs from Obama's first car, a Jeep Grand Cherokee. And he holds a special place in his heart for our nation's thirty-fifth president. One of his most prized possessions is the 1963 Lincoln Continental that took Kennedy from a Fort Worth hotel to a flight bound for Dallas on the day of his assassination. Reznikoff bought it for $17,500 and later asked for $1 million when it went up for auction. "I look at it as the car that represented the end of Camelot," Reznikoff said with measured wonderment, "because it's the last car he got out of alive."

Reznikoff had tapped into something deep. Humans, at their very core, yearn for relevance. You're nobody unless you're somebody. And for those who can't actually be somebody, they want to at least have proof that they were once in the presence of somebody. That's why people ask for autographs and to take pictures with celebrities. *Look at me. I once met Muhammad Ali. Here's a picture.* Or, more likely, *I know a guy who knows Dolly Parton's hairdresser.*

It's a basic human desire to be seen, to be counted among the many. In a sense, Reznikoff's work presaged the modern-day era of celebrity worship. Before the time of twenty-four-hour gossip sites like TMZ, Reznikoff understood the human appetite to be connected to celebrity, of relevance writ large. Thomas Jefferson once referred to the desire we attach to such things as "imaginary value." Even Reznikoff's unique obsession with famous hair was spot-on. In 2011, in an auction for charity on eBay, someone would eventually spend around $40,000 on a lock of pop star Justin Bieber's much-ballyhooed hair.

A 2011 study by a team of Yale psychologists dubbed this phenomenon "celebrity contagion." They found that fans would pay astronomical fees for a sweater worn by their idol, but only if it had *not* been dry cleaned first. "Our results suggest that physical contact with a celebrity boosts the value of an object, so people will pay extra for a guitar that Eric Clapton played, or even held in his hands," said one of the researchers, Paul Bloom. We may not ourselves have won the Nobel Prize in Physics, but touching a napkin that Einstein autographed somehow becomes a close second. At one point, someone had the bright idea to bottle and sell the supposed sweat of Elvis Presley. These star-soaked holy grails are part of our innate psychological desire to connect with the great minds of the past. I nearly tripped over myself while getting a tour of Frank Lloyd Wright's Fallingwater home outside of Pittsburgh when the docent said that, in 1939, Albert Einstein once spent the night here. (Reznikoff, of course, has the original blueprints for Fallingwater.)

As I write this book, we've reached the point on the

time-space continuum when there are fewer and fewer people who were alive at the same time as Albert Einstein. There's a famous photo from the early 1950s of Albert Einstein sitting on a chair in his backyard with two kids on his lap. They were the daughters of family friends who used to visit him in Princeton. One of those girls—Susan Sacks, now a senior citizen—lives in Israel. While visiting the Einstein Archives in Jerusalem, one of the curators asked if I wanted to spend a few hours driving out to meet her. Granted, even Sacks doesn't remember meeting Einstein. She was only three years old in the photo. But I didn't even need a moment to ponder the question: Of course, I wanted to meet her. When I got to her apartment, I kept looking her up and down like she was a visitor from another planet. I reached out and shook her hand, knowing that at some point, many decades ago, that same hand had touched Einstein. It was like a game of Six Degrees of Kevin Bacon. Yet, in this instance, there was only one person between me and Albert Einstein.

Einstein is widely considered to be the first modern-day celebrity. "He came at the moment when almost all the modern media were in place: newspapers, magazines, photographs, radio," explained Dr. Diana Kormos Buchwald, a professor of history at Caltech and an editor of *The Collected Papers of Albert Einstein*. "So, he was a quintessentially modern scientist in everything, including media exposure." In my role as Einstein's social media proxy, I wonder every day what Einstein would think of what I'm posting to his Twitter and Facebook pages.

He was the original absent-minded professor and captured the public's imagination like no other. "While lecturing on the cosmos at Caltech," writes science journalist Carolyn Abraham, he was "so engrossed in his own work that he failed to notice an earthquake that sent the rest of the campus scrambling." He personified genius, and everyone wanted a piece of it. As the culture critic Chuck Klosterman once wrote about certain artists, they "captured the zeitgeist by accidentally inventing it."

When traveling, he was stopped everywhere he went by photographers and autograph seekers. Einstein, who hated publicity, was not a fan of all the attention. An anecdote in a 1939 article about Einstein in the *New Yorker* said that when fans would approach him on the streets of New York, he would sometimes reply, "Pardon me, so sorry! I am always mistaken for Professor Einstein." He once compared autograph seekers to cannibals. "People used to eat people," Einstein said, "but now they seek symbolic pieces of them instead."

On a visit to Tokyo, thousands of Japanese fans took part in an all-night vigil outside of the hotel where Einstein and his wife were staying. When he walked out onto the balcony, the throngs cheered with rapture. Ever humble, Einstein himself couldn't grasp the appeal. "No living person deserves this sort of reception," he remarked. In 1929, when he published a paper attempting to explicate his unified field theory, one hundred journalists waited on his doorstep as if it was the birth of a royal baby. Upon the paper's publication, a department store posted it in the window to the crush of huge crowds. "I really can't understand why I have been made into a kind of idol," Einstein once said.

In 1933, fifteen thousand people swarmed into Madison Square Garden for a Hanukkah festival where Einstein was one of the honored guests. "Churches displayed statues of him," reported science writer Sam Kean. "He attended the Rose Bowl parade, and, when he visited Washington, DC, the U.S. Senate put aside its normal partisan bickering to bicker about relativity theory instead. Einstein appeared on the cover of *Time* on five occasions, and his second wife, Elsa, also got a cover story in *Time*—just for being Mrs. Einstein." In the twentieth century, people placed their faith in science, displacing religion. And Einstein was deemed a holy man, a new messiah. His quotations were cited like proverbs.

"By unraveling mysteries of the universe, Einstein had brought humankind closer to understanding God's own mysteries," wrote Abraham. Nearly seven decades since his passing, that fascination remains. "Einstein's face is the most recognizable face worldwide," said Hanoch Gutfreund, the director of the Albert Einstein Archives at Hebrew University in Jerusalem, a school the physicist helped establish. "The interest in Einstein does not fade into history. If anything, if one can say anything about this, the interest in Einstein increases with time. It's greater now."

Einstein is an icon. Quite literally. His iconic face—with the frazzled hair and the tongue sticking out—is an actual emoji you can use on your phone. Type in "Einstein" into the search bar on Amazon and you'll receive no less than ten thousand results. Sure, there's the multitude of biographies and science books. But there's also *Breakfast with Einstein, Einstein's Wife, The Other Einstein, Einstein and the Rabbi,* and something titled

Einstein on Einstein. (This book you're reading is now on that list.) Bestselling author James Patterson has written a handful of young adult novels featuring a character named Max Einstein, a modern-day teen who solves mysteries using science. And a series of a scientific primers called *What Einstein Kept Under His Hat, What Einstein Told His Barber, What Einstein Told His Cook,* and its inevitable follow-up *What Einstein Told His Cook 2: The Sequel.*

There are T-shirts and socks and mouse pads with Einstein's face on it. There is artificial intelligence software named Einstein. There are Einstein puzzles and an Einstein martini. There's an avant-garde band called Forever Einstein. They claim to be a "very smart trio." One of their albums is called "Down with Gravity." There are hospitals and medical schools named after Einstein. Britney Spears has an Einstein-inspired tattoo.

When a kid gets an A+ on his math test, his parents call him an Einstein. There was a documentary about a smart mollusk; people referred to it as "the Einstein of octopuses." To people of the modern world, Einstein is not a figure of the past. He is a living, breathing modern phenom. "The fact that he is as famous as he is, is unprecedented, and still pretty unusual," said Michael Gordin, the Princeton history professor who teaches a class about the impact of Einstein. "Children today know what his face looks like; that's kind of surprising."

For many people outside the world of science, the brand of Einstein is how they connect with him today, and his name can mean making money. In Japan, Einstein's face appears on video games. In Hungary, his picture is plastered on billboards for a local telephone company. In South Africa, he sells insurance. On a Manhattan billboard, a photoshopped image of a shirtless and buff Einstein hawks memberships to a twenty-four-hour gym. (Their slogan? "You don't have to be Einstein to see this great deal.") His quotations appear on coffee mugs. His face adorns keychains, umbrellas, frisbees, and snow globes and Christmas tree decorations. As one Hollywood agent who represents the estates of dead celebrities succinctly put it: "Einstein's the most widely recognized human being that ever lived."

This is nothing new for Einstein. "Companies he had never heard of courted him to be their spokesman after he emigrated to America in 1933," wrote Abraham. "They offered him fortunes to endorse their ties and toilet waters, their disinfectants and musical instruments, products he had never used." Einstein called these requests "a sad commentary on commercialism."

The Walt Disney Company paid Hebrew University, the caretakers of the Einstein estate, millions to name a line of popular kids' toys Baby Einstein. Apple licensed Einstein's image for its iconic "Think Different" ad campaign. From 2006 to 2017, Einstein was on *Forbes*'s list of the ten highest-earning dead celebrities. Dr. Seuss and Michael Jackson also made the list. The Einstein estate has, over the years, turned down offers for products like Einstein brain supplements. Huggies wanted to make some sort of Einstein-branded diapers. The estate politely declined.

Perhaps the most bizarre iteration of this arrived when

People magazine's 2009 "Sexiest Man Alive" issue arrived on newsstands. Actor Johnny Depp was on the cover, but it was an advertisement inside that caught the public's attention. To promote its new GMC Terrain, General Motors used a photo of Albert Einstein and put his head on a muscular man's body with an $E=mc^2$ tattoo. The slogan read: "Ideas are sexy, too." Hebrew University brought a lawsuit against the car manufacturer, claiming it didn't have a right to use Einstein to sell a sport utility vehicle. They added that "Dr. Einstein with his underpants on display" causes damage to the "carefully guarded rights in the image and likeness of the famous scientist." But GM argued that the "advertisement does not state expressly (or even imply) that Dr. Einstein (or Hebrew University) endorsed the Terrain, nor would any reasonable reader reach that conclusion." For good measure, the company added that the whole endeavor was "to make a light-hearted point about the smart (but sexy) features of the Terrain." Why anyone would apply sexual terms to a two-ton piece of metal is beyond me. Regardless, Judge A. Howard Matz, a federal judge in California, ruled in favor of GM. The university appealed and the two parties eventually settled out of court.

No less strange is the case of Einstein Bros. Bagels, a national chain named after the company's fictional founding brothers, Melvin and Elmo. A 2019 article in the *Washington Post* revealed that the wealthy German family that owns the chain had strong ties to the Third Reich. "It is all correct," said a family spokesman, referring to the findings that, during World War II, they had donated money to the Nazis. A smidge ironic, considering Einstein himself denounced his German citizenship as Hitler came to power. Or to quote a tweet from when the news was announced: "Is anyone really surprised though, their bagels are 100 percent a crime against the Jewish people."

Einstein today has become a brand, a totem—in many ways bigger than the man he actually was. What the *New York Times* TV critic James Poniewozik wrote about Donald Trump in his 2019 book *Audience of One* could just as accurately be said about Einstein: "The best and worst leaders build themselves out of stories. They use their culture's language—legend, metaphor, archetype—to express what literal language can't… He's a character that wrote itself, a brand mascot that jumped off the cereal box and entered the world, a simulacrum that replaced the thing it represented."

One hundred years ago, few people outside Germany had ever heard of Albert Einstein. But today he's a celebrity with a massive online following. He has more Facebook fans than the Rolling Stones. "Einstein is a cool guy with the tongue, with the hair," Roni Grosz, the curator of the Albert Einstein Archives told me when I visited him in Israel. "But come on—tongue and hair, we've seen that with others. We have Mick Jagger. But Mick Jagger's not on par with Einstein."

EINSTEIN AND TIME TRAVEL

"The distinction between past, present, and
future is only a stubbornly persistent illusion."

—ALBERT EINSTEIN

was first introduced to time travel when I was ten years old.
I remember it like it was yesterday. It was mid-July 1985, the
summer between fourth and fifth grade. My older brother
Dani, who had recently got his driver's license, took me to the
movie theater to see *Back to the Future*. And then I made him
take me again, and again.

I assume nearly everyone has seen the movie by now, so I
don't feel like I'm spoiling anything by giving away the major plot
points: There's a scientist named Doc Brown who has somehow
managed to convert a DeLorean into a time machine. Portrayed by
actor Christopher Lloyd, Brown is a doppelgänger for Einstein if
there ever was one—embodying the ethos of a mad scientist with
unkempt and wild white hair. "Albert Einstein served as one of
my greatest inspirations for Doc Brown," Lloyd said. (It turns out
that Lloyd is actually distantly related to the Nobel Prize–winning
physicist. My friend AJ Jacobs wrote a book called *It's All Relative*

about genealogy and finding lost relatives using online services like 23andMe and MyHeritage. I asked him to run a search for me, and here's what he discovered: Albert Einstein is Christopher Lloyd's aunt's first cousin's husband's first cousin's husband's uncle's wife's first cousin's husband's nephew. So, you know, there's that.)

In the film, Brown explained how he was standing on a toilet hanging a clock. The porcelain was wet, he slipped and hit his head on the edge of the sink. "When I came to, I had a revelation, a vision, a picture in my head, a picture of this," he said while pointing to the Flux Capacitor, the secret sauce that turned a sports car into a time-traveling utility vehicle.

"I used to read about scientists like Einstein," Lloyd said. "People get so involved in something, and then they're getting on a bus and—bingo!—$E=mc^2$," he laughed. "And I felt that Doc had a lot of that in him."

As if to further cement the connection, Brown's pet is a Catalan sheepdog with white, scruffy hair named Einstein. Einstein is the movie's first time traveler, jumping ahead one minute into the future during a test run of the machine. The original version of the movie featured a pet monkey named Shemp, but that idea was ultimately nixed because Sid Sheinberg, an executive at Universal Studios, felt that movies with monkeys in them didn't fare well at the box office. (I guess he hadn't heard of Planet of the Apes.)

Anyway, back to the movie: Through a series of events that involve Libyan terrorists looking for plutonium in a mall parking lot (ah, to be a scriptwriter in the 1980s), the DeLorean gets put to the test. Its first human occupant? Marty McFly, Brown's young friend played by Michael J. Fox, and he's headed to 1955. That's where he bumps into his parents, who are in high school, and he unwittingly interrupts their courtship. Marty's presence in their lives accidentally changes the course of his family's history, and he spends the remainder of the movie trying to make sure his mom and dad attend the high school dance together, have their first kiss, and fall in love—thereby ensuring Marty's very existence in the future.

The Doc Brown character left an indelible mark not only on Lloyd, but also on the millions of people who saw the movie, and its subsequent sequels. "So many people have come up to me and told me how the film made their childhood," Lloyd recalled. "Or that they became engineers or scientists or surgeons or whatever from the effect of the film. And nothing else I've done has had that kind of impact."

I was one of those kids. For me, the film changed the aperture of how I saw the world. It had action, comedy, romance. I was hooked. Up until that point, I had seen movies that took place in reality: *The Karate Kid*, *The Black Stallion*, *Kramer vs. Kramer*. (Yes, my parents took me to see *Kramer vs. Kramer*. I guess they couldn't find a babysitter that day.)

But *Back to the Future* was something entirely different. It bent genres and colored outside the lines. It showed me that anything was possible, and it became my life's guiding principle. I began writing fantasy novels (none published, although, fingers crossed, they still exist on a floppy disk in a dust-covered box in my attic), I launched a high school newspaper, and taught myself to skateboard so I could be like Marty in the movie. The shortest guy in school, I somehow wiggled my way onto the basketball team. Like I said, anything was possible.

I have since become obsessed with time-travel movies. Wikipedia lists at least four hundred of them in an entry accurately titled "Films about Time Travel." (Fifty of them are different adaptations of *A Christmas Carol* with its ghosts from Christmas past and future. Another eleven are based on Mark Twain's *A Connecticut Yankee in King Arthur's Court*. The online encyclopedia helpfully acknowledges the caveat that this category is constantly changing with time. But I digress.) While none carry the same nostalgia and fondness of *Back to the Future*, I'm always game to rewatch *Time Bandits* and *Groundhog Day*. *About Time*, about a man who tries to fix the past to improve his present, is my favorite romantic comedy. And don't get me started on TV shows. When NBC was thinking about canceling *Timeless*, a series about a ragtag team that travels back in time to stop an evil corporation from changing history, I was one of the many die-hard fans who were part of a campaign to bring back the show. (NBC acquiesced.)

All of this—the DeLorean, the skateboarding, my pitiful plea to a major TV network—fueled a lifetime of fascination with time travel. And then one day I was scrolling through news headlines—climate change, war in Ukraine, Kim Kardashian— and then this stopped me in my tracks: a story about a physics professor named Dr. Ronald Mallett. This guy had invented the world's first time machine.

Driving to Mallett's home in Coventry, Connecticut, one summer afternoon, I found myself in a town not unlike the fictional Hill Valley of *Back to the Future*. I drove past Coventry High School, past the Coventry Volunteer Firefighters Association, past the quaint downtown shops festooned with American flags. When I pulled up to his three-story condo, Dr. Mallett stood outside waving to greet me.

Mallett is six feet tall with a thin pencil mustache and black hair specked with silver. He's dressed in a light blue button-down shirt, dress pants, and gray slippers. (Mallett is obsessed with Einstein, and I could only guess this was a nod to Einstein's real-life penchant for furry footwear.) He appeared just as he did in the countless TV appearances I watched of him. There he was on CNN, on NBC News, and in not one, but two documentaries about his work.

His home office is lined with filing cabinets, a bust of Einstein rests near his cluttered desk. A clock featuring Marilyn Monroe hangs on the wall. In fact, his entire home is decorated with art related to either Einstein or Monroe. Behind his desk was a painting made by a friend: it features Einstein and Monroe, with Mallett standing in between them. (A cousin once joked that the only person Mallett would marry would be someone "as brilliant as Einstein and as beautiful as Marilyn Monroe.")

"She started out with nothing," he said, when I asked what drew him to Monroe. "And she was able to achieve so much with her life. This drive to go beyond where she was, was something that mesmerized me." He paused. "Plus, you know, she was beautiful."

Mallett isn't the only person who envisioned those two together. A play called *Insignificance* (later adapted into a movie) follows four famous icons of the twentieth century who converge at a New York City hotel one night in 1954: Joe DiMaggio,

Joseph McCarthy, Albert Einstein, and Marilyn Monroe. There's even a scene where the bombshell attempts to breathlessly explain the special theory of relativity.

Much of Mallett's office is itself frozen in time: A boom box (a boom box?!) collects dust on a shelf. On a bookcase, as if from a pre-streaming era, is a collection of DVDs: *Casablanca*, *Star Trek*, *The Iceman Cometh*, and, of course, *The Time Machine*. Worn, decades-old physics textbooks crowd other shelves. "Books give me a sense of peace," he said, as if his quest for knowledge needed an explanation.

Perhaps the one sign that we are actually in the twenty-first century is his computer, which has a special keyboard used by hard-core video gamers. I asked what his favorite game is.

"*Assassin's Creed*," he said, referring to the popular series based on a Viking tale. "I like adventure games that actually have some sort of epic story associated with it." He preferred games that have an immersive world, like time travel in which he can transport himself from Connecticut to ancient Egypt.

Does he play against other people over the internet?

"No, no, no. I'm a loner. By core and by choice, I'm a loner," he said. "Before my father's death, I was pretty gregarious. But after his death, I became very solitary." His dad's death sparked an obsession within Mallett: He became singularly focused on figuring out a way to time travel, to go back to the past, and save his dad.

Mallett thought his father, Boyd, was a superhero, someone who was immortal. But time stopped for the family on May 22, 1955.

That's when Boyd died from a massive heart attack after a party on the night of his eleventh wedding anniversary, leaving behind a thirty-year-old widow and four kids. Ron, the oldest, was just ten years old.

"It shattered my world," Mallett recalled. He went into a deep spiral after his dad died. He became depressed and would skip school; he immersed himself in science fiction books in his room. Anything to transport him somewhere else where he could dull the pain.

He was a fan of Classics Illustrated, a series of comics based on famous books. He spent his weekly allowance amassing a large collection: *Knights of the Round Table, 20,000 Leagues Under the Sea, The Iliad*. Then one day, while perusing the racks at the drugstore near his home, he discovered the issue that would change the course of his life: H. G. Wells's *The Time Machine*, originally published in 1895.

The story—of an inventor in Victorian England who travels into the distant future—showed Mallett that time is merely space, and that we could move forward and backward in time just as we could move forward and backward in space. It concretized an amorphous concept. "When I read that, that was like a life preserver to me," Mallett recalled. "Because I thought, if I could build a time machine, then I could go back in time and see my father again and tell him what was going to happen." He said he would tell him to eat healthier and stop smoking. "Maybe that would save his life. And so that literally became my mission."

When the 1960 movie version of *The Time Machine* arrived at the local theater, Ron saw it five times. For the next hour,

Mallett unspooled the story of his life, gliding over his background, flicking in bits of introspection.

His father had been a TV repairman in New York—fixing sets so people could watch *The Lone Ranger* and *I Love Lucy*. Occasionally, Boyd found himself repairing the television in the home of a bona fide celebrity like Jackie Cooper or Walter Matthau. Mallett would wait by the subway stop each evening for his father to return from work so he could carry his dad's toolkit on the way home. "There was just this aura around him," Mallett recalled. "I was like a prince around the king."

Mallett went to the basement and dug up the tools from his dad's TV repair kit. He found discarded wires, bicycle parts, and vacuum tubes, and tried desperately to recreate the time machine he saw on the cover of the comic book. The pièce de résistance was a gyroscope that Mallett's father had given him, much in the same way Einstein's father gave his curious son a compass. When Mallett thought he was done cobbling it all together, Frankenstein-like, he crossed his fingers and plugged the cord into the wall. When it didn't work, eleven-year-old Ron came to a stark realization: He couldn't learn how to successfully travel through time from a comic book. "I wasn't discouraged," he said. "I knew my only limitation was that I wasn't a scientist."

He became so enamored with the actual mechanics behind time travel that he siphoned off his lunch money so he could buy used science books at the local Salvation Army store. "I would literally rather read than eat," he told me. (He got so thin from the lack of eating that the family doctor assumed Ron had an iron deficiency. His mother started packing him a lunch from home instead.) It was at that bookstore that he stumbled across the second book that would change his life: it had on its cover a drawing of Einstein next to an hourglass.

Ron had heard of the great genius. Einstein had passed away just a month before Ron's dad died. The physicist's passing had been all over the news and now here was this book, *The Universe and Dr. Einstein*, reaching out to Mallett. He read it voraciously cover to cover and, while he didn't understand it completely, he did learn one very important thing. According to Einstein and his colleague Nathan Rosen, an Israeli physicist, we have the ability to travel through time, thanks to tunnels known as wormholes. Google "wormholes" and you'll see that scientists refer to them as Einstein–Rosen bridges, perhaps the most Jewish-sounding overpass known to man.

Motivated by his dad's work as a TV technician, Mallett signed up for an electronics class in his junior year of high school where he was taught the basics of wiring and circuitry. He wanted to become an electrical engineer. He hatched a seemingly ridiculous plan: he was going to build this time machine. "It became my total obsession."

But to get from here to there, from a pipe dream to possibly reuniting with his father, he would need to learn more: physics, electrodynamics, quantum mechanics, and, of course, the theory of relativity. That seminal paper by Einstein showed that time and space are malleable and that, if conditions are just right, a wormhole could be created that would allow us to travel through time.

Unable to afford college, Mallett took advantage of the GI

Bill, joining the Air Force, where he helped maintain its system of nationwide computers for the Strategic Air Command command-and-control centers. Eventually, he was accepted into the physics program at Penn State.

In school, he kept his quixotic passion close to his vest. After all, how would it sound if he walked up to his professors on the first day of classes and told them he wanted to build a working time machine. Instead, he chose to focus his studies on another area of Einsteinian physics. "Black holes were considered to be a crazy idea, but legitimately crazy as opposed to time travel," Mallett explained. "And black holes do affect time. So, I knew that if I studied black holes, that became my cover story, literally."

He received his doctorate from Penn State University in 1973 after only two-and-a-half years of study. At the time, he was one of only seventy-nine African Americans with a PhD in physics. He took a gig in the research department at United Technologies, a multinational conglomerate that built everything from aircraft engines to escalators. The salary allowed him to live comfortably, but it was not his dream job in academia. "I took heart in remembering that even Einstein had been unable to find a teaching job upon graduating from college and had been required to take a position in the patent office to pay the bills."

He eventually got a professorship at the University of Connecticut, but as the years went by and he was still no closer to building a time machine, a depression washed over him. Each night he would come home after work, ignore his wife, and go straight up to the bedroom where he would stare at the ceiling and listen to sad songs by Simon & Garfunkel. "You have to remember, my identity was wrapped up in this thing. If I don't have this goal of the time machine, then what do I have?"

He got divorced, had his own scare with a near heart attack, and redoubled his efforts—once and for all—to see the time machine through to its completion.

And then something remarkable happened. He built it.

Ronald Mallett informed the public that he had constructed a mathematical model—the blueprints, if you will—of the world's first time machine in April 2000, with a scientific paper called "Weak gravitational field of the electromagnetic radiation in a ring laser." With a name like that, it—not surprisingly—didn't attract all that much media attention. Months went by: George W. Bush became the forty-third president of the United States and *Dude, Where's My Car?* arrived in theaters—both to mixed reviews.

But everything changed when the May 19, 2001, issue of *New Scientist* magazine hit the newsstands. The editors had decided to devote the lead story to Mallett and his research. "Presenting the world's first time machine," blared the tantalizing headline on the cover. The rest, as they say, is history.

Mallett was no longer working in obscurity. He was fawned over in the *Wall Street Journal*, the *Boston Globe*, and *Rolling Stone* magazine. Documentary crews were knocking on his door. He was receiving so many emails, there was no way he could respond to them all. It's at this point, after an hour of hanging out with Mallett at his house that I ask him to please, for the love of God, tell me how his concept of time travel works. He said that his calculations are based on two of Einstein's seminal theories.

The essence of the special theory of relativity, published during Einstein's "Miracle Year" of 1905, is that time is affected by speed. He told me about an influential experiment conducted in the autumn of 1971. Two scientists—physicist Joseph C. Hafele and astronomer Richard Keating—boarded Pan Am Flight 106 from Washington, DC, to London. Resting on the seats next to them were two atomic clocks, the most precise timekeeping mechanisms that exist. Their goal was to test Einstein's special theory of relativity—in particular his "clock paradox," a theoretical prediction that claims that time moves more slowly for a clock aboard a rapidly moving object like a commercial jet, than a clock that's hanging on your wall at home. In essence, Einstein said that time is different inside a speeding plane than in a plane that's parked on the tarmac.

They received $8,000 in funding for the project—$7,600 of which they spent on four round-the-globe plane tickets, two for themselves and two seats on each flight for "Mr. Clock." While they were in the air, colleagues back on the ground were monitoring the atomic clocks in DC at the U.S. Naval Observatory. When the scientists landed, the clocks aboard the plane were slightly behind the clocks that were left on the ground.

What they found is exactly what Einstein predicted: that moving objects age slower than stationary ones. "So those scientists that were on board actually aged less than their colleagues who were stationary," Mallett said, a sense of wonder coming across his face. The clocks in the air and the clocks on the ground may have shifted by just a few seconds but, Mallett explained, "eventually, when we have rockets that can go fast enough, we will see this effect and it will have sociological implications. Because if an astronaut travels close to the speed of light, when they come back, only a few years may have passed for them. But decades may have passed for their family and everyone else. They could come back and find out that they're actually younger than their children. And this is time travel into the future."

As of the printing of this book, Russian cosmonaut Gennady Padalka holds the record for the person who has spent the most time in space—at 879 days. Do the math, and that means he's about one second younger than he would be if he hadn't boarded a rocket ship. He is our generation's greatest time traveler.

These kinds of experiments are occurring on a regular basis in an underground tunnel that spans seventeen miles beneath the France–Switzerland border. That's where you can find the Large Hadron Collider, the world's largest particle accelerator. It was built in 2008 with the help of ten thousand scientists, and it uses an effect called "time dilation," discussed in Einstein's special theory of relativity, to routinely send subatomic particles into the future. (Some people blamed a 2016 earthquake in Italy on the machine.)

While time travel to the future has been proven possible based on the special theory of relativity, it doesn't allow for travel to the past. However, there is another way, and that's where another theory of Einstein's comes into play. The core of the general theory of relativity, published in 1915, is that time is affected by gravity. If you could find a place where gravity could

be twisted—like inside a rotating black hole–you could theoretically twist time. Picture it like this: Find a thin strip of paper.
On the bottom write "PAST," in the middle write "PRESENT,"
and on the top write "FUTURE." If this piece of paper is lying on
your desk, it looks like a simple timeline: past, present, future.
But what if there was some way to take that flat strip of paper and
turn it into a bracelet? What Mallett was explaining was that if
this strip of paper got tossed into a black hole—or another place
where gravity was twisted—then the paper could be turned into
a loop where you could travel from the present to the past to the
future and on and on. The timeline is no longer flat, but round.

So, is there a way to twist gravity that doesn't involve flying
to space and finding a black hole to jump into? Einstein's general
theory of relativity posits that light can impact gravity. "This is
the connection that led to my breakthrough," Mallett said. "In
other words, what I realized was that if gravity can affect time
and light can create gravity, then light can affect time."

Indeed, in 1949, the famous Austrian mathematician Kurt
Godel published a paper showing how this was possible. Einstein
himself commented on Godel's work, saying that it "constitutes,
in my opinion, an important contribution to the general theory
of relativity, and especially to the analysis of the concept of time."
Wormholes, cosmic strings, rotating black holes, and rotating uni-
verses are all the result of Einstein's general theory of relativity and
show that time travel to the past is possible. The work of Godel
and other physicists—namely Brandon Carter, Richard Gott, and
Kip Thorne—laid the foundation for Mallett's discovery.

Mallett quickly got up from his chair and went into the

kitchen to make a cup of coffee. Moments later, he carried the cup to the table where I was sitting. "Imagine that the coffee in this cup was like empty space. Now imagine," he said, as he grabbed a spoon and stirred the coffee, "that this spoon is a light beam. It's actually causing empty space to swirl around. By twisting space, I can twist time, and by twisting time I can travel back into the past."

But how could he prove this outside of a coffee cup? Mallett said when he was working through these equations, he couldn't sleep. He spent days scribbling on notepads in his den while the music of Wagner's operatic masterpiece *The Ring of Nibelung* emanated from the record player. After several weeks, and checking and rechecking his math, he finally figured it out.

It's important to point out that Mallett is a theoretical physicist. In physics, there is a division of labor between theoretical physics and experimental physics. Broadly speaking, theoretical physicists, like Einstein, use mathematics to try to explain how the universe works, while experimental physicists turn those equations into something physical. So Mallett enlisted the help of a colleague, an experimental physicist named Chandra Roychoudhuri.

What they built is a model time machine, a physical representation of Mallett's mathematical equations. It's the size of a blender, a small glass cylinder equipped with lasers set up to create a circulating light beam. (Scientists refer to it as a ring laser. Think of it as Mallett's own version of a Flux Capacitor.) He flipped on the lasers, and a circulating light beam was produced. If the beam is intense enough, it could warp the empty space inside the blender: twist space and twist time. "What I had discovered was a time machine, essentially based on light."

The logical next step for Mallett's research is to take what he's discovered in a laboratory and put it to the test. "Unlike Doc Brown in *Back to the Future*, this is not something I'm going to be able to put together in my garage," he said, reminding me of the *theoretical* in the theoretical physicist line of his resume. He estimated this would cost in the neighborhood of $250,000. That dollar figure sounds ludicrously low for the ability to go back in time, to stop Lincoln from being assassinated or to know who would win the World Series—but that's just for a feasibility study. The cost of full-scale experiments could run into the millions or billions of dollars. The funding would probably require the support of major foundations or national governments or some very wealthy individuals. I can imagine Elon Musk or Vladimir Putin willing to pay ten times that amount for a working time machine.

Mallett agreed. "If the CIA found out that North Korea was investing in time travel, I would probably have more money than I knew what to do with."

So, what's preventing them from investing in his research? "They don't know about me," he said, adding that it's one of the reasons he invited me over today, to interview him and devote a chapter of my book to his work. Because, you know, Russian dictators are exactly my target audience.

"This is not an uncommon strategy," he told me. The scientists behind the atomic bomb were worried that the Nazis were already working on an atomic weapon. So, they tried to get the attention of President Roosevelt. What did they do? They got

Einstein—the most famous scientist in the world—to sign on to a letter. "That got Roosevelt's attention."

I could relate with Mallett's desire to reunite with a deceased parent. My own life careened off course when my mother died suddenly when I was in eighth grade. She was alive when I left the house to go to school. Later that morning, as I sat in my American history class learning about Shays' Rebellion, she had a fatal brain aneurysm. Months earlier, at my bar mitzvah, I had become a man in the eyes of my faith. And now, well, now I was curled up in the fetal position. Never did I wish for a time machine more than at that precise moment.

What I wouldn't give to hop into a DeLorean and see my mother one more time, to have one final conversation where I warned her to go to the doctor for a checkup. Isn't time travel a superpower we all desire? There are moments when we all wish we could hit an "undo" button in life, like we do on a computer. Just said something stupid at work? No problem. Just hop into the ring laser blender and go back five minutes to before you embarrassed yourself in front of your boss.

In the Jewish faith, children mourn for a year after a parent's death. Under the strictest interpretation of that law, it means not going to parties, concerts, or even to the movies during that period. *Back to the Future II* was released eleven months after my mom's funeral, which meant I couldn't see it right away. Thankfully, my best friend Chaim waited to go, out of solidarity, until my year of mourning was over. It was the first thing I saw.

In the movie, Marty travels ahead to 2015 to see how his own future turns out. For my fortieth birthday, which also took place in 2015, my wife threw me a surprise *Back to the Future*–themed party. Everyone arrived in costume as characters from the film. I, of course, was dressed as Marty. My wife, in an Einstein wig, was Doc Brown. As I type this, I'm literally wearing a *Back to the Future* T-shirt while, in my office, I'm surrounded by artwork and knickknacks from the movie. Yes, I have the schematics for the Flux Capacitor framed on my wall.

If my wife is reading this—and that's a big *if* considering she's not always interested in what I write—all I want for my fiftieth birthday is a ride in a DeLorean.

While Mallett is certainly a big fan of *Back to the Future*, he told me that his favorite time-traveling movie is actually *Timecop*. The 1994 film is set in a world where time travel is a regular occurrence. It stars Jean-Claude Van Damme as a member of the Time Enforcement Commission, a group that was created to police time travel, to ensure that nobody goes back in time to game the stock market, for example. Criminals are charged with "time travel with intent to change the future."

What draws Mallett to rewatch *Timecop* over and over is that it looks at time travel through an ethical lens. There are what he calls "temporal historians" whose job it is to prevent people from changing the way things were supposed to occur. "Any technology has a double edge associated with it," he admitted. "It can be used for good or bad. Think about a butter knife. It can be used to spread butter, but it can also be used for other things." I think I know what he's trying to say, but now all I can think about is the

Zodiac Killer stabbing his victims with a tiny, and not too sharp, kitchen utensil.

As Paul Virilio, the French philosopher, once noted: "When you invent the ship, you also invent the shipwreck; when you invent the plane, you also invent the plane crash; and when you invent electricity, you invent electrocution... Every technology carries its own negativity, which is invented at the same time as technical progress."

If you happened to be reading an obscure online discussion board on November 2, 2000, you may have noticed a strange message appear: "Greetings. I am a time traveler from the year 2036."

The poster, who went by the name John Titor, weaved a remarkable tale. He claimed to be a soldier, part of a government-run time travel operation, based in a futuristic Tampa, Florida. He had been sent back to 1975 to grab an IBM 5100 computer the military desperately needed to help them debug something in the future. Because, of course, it makes total sense that the fate of all humanity is somehow connected to an Elvis-era computer. Titor said he had taken a detour and stopped over in the year 2000 for "personal reasons" to visit family.

For the next year, the posts continued, each with more detail about the operation. Titor said the time machine was installed in the rear of a 1967 convertible Chevy Corvette, and he provided schematics for all to see. (Online sleuths said the engine looked more like the one in a 1987 Chevy Suburban.) He talked about an impending civil war, to start in 2005, with the United States breaking into five parts. This would be followed by World War III, after which Omaha, Nebraska, of all places, would be the nation's new capital. In 2015, he claimed, three billion people would die because of a nuclear war.

After the attacks of September 11, 2001—an event which Titor, regrettably, did not warn us about—some believers started to disperse. After all, Ben Stiller's *Zoolander* had just been released, and that movie wasn't going to watch itself.

Titor tried desperately to maintain his relevance, first warning people like a Biblical prophet—"Learn to shoot and clean a gun. Consider what you would bring with you if you had to leave your home in ten minutes and never return"—and then just being an all-around jerk: "Perhaps I should let you all in on a little secret. No one likes you in the future. This time period is looked at as being full of lazy, self-centered, civically ignorant sheep. Perhaps you should be less concerned about me and more concerned about that."

At one point, Titor brought up the possibility of bringing volunteers with him to the future. "For all of you interested in coming back with me to 2036, perhaps we should discuss the trip," he wrote, as if he was inviting buddies to join him on a fishing vacation. Alas, nobody that we know of took up Titor on his offer.

When someone asked him who would win the Super Bowl, Titor was petulant. "If a time traveler had knowledge of your future, and you could only ask one question, would this be it? Besides, can you tell me if it rained in New York on June 4, 1932?"

When his predictions didn't come true—he said, for example, there would be no more Olympic Games after 2004—hard-core Titorists came to his defense. They brought up the "many worlds" hypothesis in quantum physics, that the universe has potentially more than one timeline. And, besides, Titor's presence in our present may have impacted future events. An ontological paradox known as the Novikov self-consistency principle, named after Russian astrophysicist Igor Dmitrievich Novikov, states that any actions taken by a time traveler while time traveling were part of history all along. (This concept is charmingly employed in the 1989 movie *Bill & Ted's Excellent Adventure*, when a present-day character is looking for a lost set of keys and then wonders, "What if I went back in time and put the keys right over there?" and then proceeds to find the missing keys exactly where his hypothetical self had just placed them.)

There are, of course, enough time-travel paradoxes that you could print them all out, double-sided, and you wouldn't have room to stuff them all inside a DeLorean. Take, for example, a global pandemic. "Say you traveled in time, in an attempt to stop COVID-19's patient zero from being exposed to the virus," explained Dr. Fabio Costa, a physicist in Australia. "However, if you stopped that individual from becoming infected—that would eliminate the motivation for you to go back and stop the pandemic in the first place." You could also call this the "Killing Hitler" paradox: go back in time and kill baby Adolf, but then nobody would know about Hitler, and you'd have no reason to go back in time in the first place.

Costa and his colleague, Germain Tobar, claim to have figured out a mathematical solution to this problem. "In the coronavirus patient zero example, you might try and stop patient zero from becoming infected, but in doing so you would catch the virus and become patient zero, or someone else would," Tobar said. "No matter what you did, the salient events would just recalibrate around you. This would mean that—no matter your actions—the pandemic would occur, giving your younger self the motivation to go back and stop it."

A Google search for "John Titor time traveler" delivers more than fifty thousand results. One of those was for an academic article in a science journal written by Alasdair Richmond, a senior lecturer in philosophy at the University of Edinburgh. If anyone would want to time travel and warn someone of impending danger, it would be him. When Richmond was twelve years old, his father was literally killed by a bolt of lightning. But, unlike Mallett, Richmond is under no pretense that he could go back in time and warn his dad of impending doom. "I think changing past events has precisely zero probability," he wrote.

Richmond is working on a book called *Time Travel for Philosophers* and has a YouTube channel where he discusses the intricacies of time travel in his Scottish accent and dressed like a professor at a steampunk-themed Halloween party.

It was Stephen Hawking who added yet another nail in the coffin of the Titor hoax. On June 28, 2009, the celebrity physicist tried an experiment. He hosted a cocktail reception for time travelers at the University of Cambridge. The catch? He didn't reveal the invitation until *after* the event. "Maybe one day," he explained, "someone living in the future will find the information and use a wormhole time machine to come back to my party, proving that time travel will one day be possible."

Alas, nobody showed up to his party. There were many online theories to help explain how time travel could still exist even though nobody had the good sense to attend a party hosted by Stephen Hawking. One article titled "Five Possible Reasons Nobody Showed Up for Stephen Hawking's Retroactive Time-Traveler Party" offered one simple, non-scientific explanation: "Time travelers are dicks."

So, what was Hawking's response when nobody arrived? "What a shame," he said. "I was hoping a future Miss Universe was going to step through the door."

Two physics students at Michigan Technological University conducted an experiment in 2013 to finally, once and for all, determine if time travelers were visiting us. Robert Nemiroff and Teresa Wilson came up with three tests using so-called "digital signatures."

The first thing they did was to look at online posts from January 2006 to September 2012 to see if anyone had looked for either the term "Pope Francis" or "Comet ISON." Both terms were unique—nobody had ever been called Pope Francis before Jorge Mario Bergoglio took the name in 2013, and ISON, discovered in late 2012, is the only comet to have that moniker. They wanted to see if anyone had accidentally used those terms before they existed.

Second, they ran a report of Google searches from that time period. "A time traveler might have been trying to collect historical information that did not survive into the future or might have searched for a prescient term because they erroneously thought that a given event had already occurred, or searched to see whether a given event was yet to occur," Nemiroff and Wilson wrote.

Third, they created two never-before-used hashtags—#ICanChangeThePast2 and #ICannotChangeThePast2—and asked any time traveler to use one of those in a tweet before August 2013. Nobody did. The students pointed out that the negative results don't disprove time travel—it's quite possible that not everyone from the future uses Twitter.

It's at this point in time when Mallett finally admitted to me that there's a small wrinkle in his plan. Using Einstein's theories to build his time machine comes with a caveat. The furthest he can travel back in time is to the exact moment when the ring laser machine was first turned on. Which means, since he made this discovery long after his father passed away, he wouldn't be able to use it to go back and see him. At best, he might be able to attend a Britney Spears concert, circa 2003. But Mallett, ever the optimist, has a solution.

"Let's suppose that eventually we are able to develop the technology to go to an advanced planet. And suppose that they've had a time machine turned on for ten thousand years, right? You can see where I'm going?"

I do. He's saying that it's possible aliens or another civilization have turned on a ring laser time machine many, many years ago. All he'd have to do is ask them to borrow it so he could visit his dad before the heart attack. Sure, it's a stretch, but it's theoretically possible.

"Who knows?" Mallett asked. "Because to me, the universe is so vast with so many possibilities. Can I rule that out? No."

In an alternate universe, there's a sign that says MALLET & SONS TV REPAIR. His father had been grooming him to join the family business. "The whole arc of my life would be different," Mallett said. But now, he is once again alone. His singular obsession with time travel has wreaked havoc on his personal life. He's been married three times and has no children. He told me I was the first visitor he's had over in months.

Asked what he would say to Einstein if he went back in time to meet him, Mallett replied: "The two most important men in my life were Albert Einstein and my father. I would thank Einstein for the work that he's done and for having given a poor ten-year-old hope for his life. Without Einstein, I would've felt hopeless, I would have felt lost."

Mallett once gave a presentation of his work to an audience of relativity specialists at a science conference, and he spoke about his personal motivation with the death of his father for sparking his interest in time travel. In the audience was the highly respected theoretical physicist Bryce DeWitt, who was the director of the Center for Relativity at the University of Texas in Austin. During the question-and-answer period, DeWitt got up and made a comment to the entire audience that

Mallett still remembered. "I don't know if you'll see your father again," DeWitt said, "but he would have been proud of you." For Mallett, DeWitt's words made his lifelong struggle worthwhile.

EINSTEIN LIFE HACKS

"I have no special talents. I am
only passionately curious."

—ALBERT EINSTEIN

Albert Einstein didn't like to wear clothes.

To be clear, he just didn't like to get dressed up. Why not, you know, be comfortable? So, while everyone else zigged, he not only zagged, but discarded social graces at the door and then, well, took off down the street in the same clothes he slept in. It was a common sight in Princeton: the Nobel Prize winner in flannel pajamas, shuffling around his neighborhood like a Jewish Hugh Hefner.

Einstein's distaste for fashion included footwear. His big toe kept making holes in his socks so, at some point, he just gave up on socks altogether. Besides, Einstein was deemed unfit for the Swiss Army on account of, among other things, varicose veins and sweaty feet. "When he couldn't find his sandals, he'd wear [his wife's] sling backs," the BBC reported. I shared a photo of Einstein wearing his wife's sandals on Facebook and it garnered 150,000 likes in just a few hours. Science writer Carolyn

Abraham, who wrote a book about Einstein, called him a "galactic clairvoyant, grounded to Earth and lesser mortals by his shabby sweaters and sockless feet."

He wore sweatshirts that grew rattier over the years because wool sweaters made him itch. When his wife, Elsa, "urged him repeatedly to dress up for important company," Einstein said, "that if the important visitors were coming to see him, they'd see him as he was; but if they were coming to see his clothes, Elsa would show them to his wardrobe."

In my quest to discover all the ways in which Einstein is relevant today, I've decided I need to explore his daily life and see what lessons I could learn. I may not fully comprehend the equivalence principle, but only dressing in comfortable outfits? I could get on board with that. What else could Einstein the man, not Einstein the scientist, teach us? What life hacks could I incorporate into my own daily existence?

"It is almost as if the discoverer of new laws confirming some verifiable intentions within creation thereby come closer to the creator himself," psychologist Erik H. Erikson wrote about Einstein. But he was far from godlike. "He was slovenly," said Robert Schulmann, a former editor of the *Collected Papers of Einstein*. "And at some point, it began to work in his favor." I could do slovenly. Maybe not comb my hair, embrace a shaggy and rumpled appearance. And so, I began with the pajamas. I sleep in a T-shirt and sweats. Each morning after I showered, I simply put on a new pair of pajamas. My wife wasn't too thrilled, but our pets didn't seem to mind as long as I still gave them treats and, fortunately, I have a job where I work from home. To be honest, fashion was never really at the top of my priority list anyway.

I wore my pajamas to the grocery store and to my dentist appointment. I gurgled and made a mess, but it didn't matter. Maybe Einstein was onto something. I even showed up to vote in a local election looking like I had just rolled out of bed. Democracy in action, sans the pants. But like Einstein, I live in a college town. Half of our population are students who attend class wearing pajamas. My sartorial choice may have been an oddity in a big city, but here I merely blended into the background. Nobody even gave me a second glance.

In his later years, Einstein adopted somewhat of a uniform: a signature gray suit with no socks. Most of the time, he'd nix the fancy sport coat and wear his leather Levi's Menlo Cossack jacket. Similarly, Apple co-founder Steve Jobs became famous for not deviating from a black shirt and jeans. Ditto for Facebook CEO Mark Zuckerberg. "Uniform dressing has roots in not just physical but mental efficiency," writes fashion journalist Elyssa Goodman. When Einstein was employed by the patent office, he was able to do a full day's worth of work in only two or three hours. "People who have to make immense decisions every day will sometimes choose a consistent ensemble because it allows them to avoid decision fatigue, where making too many unrelated decisions can actually cause one's productivity to fail." In this way, Goodman says, Einstein might be the forefather of "normcore," a modern aesthetic characterized by unpretentious, average-looking clothing. I found a sweatshirt and sweatpants that I liked and ordered ten pairs online. I'm not sure if the ensemble made me more productive at work, but at least I felt like I was being enveloped by a soft blanket 24/7.

In my research, I found that Einstein didn't like going to the barber because it took too much time away from his studies. So, his wife, Elsa, who happened to be nearsighted, cut his hair. "She was, so to speak, flying blind and the haircut turned out accordingly," recalled Herta Waldow, the Einsteins' live-in maid from 1927 to 1932. "There exist photos in which Einstein's head looks like a plucked chicken." So I purchased a haircutting kit on Amazon and asked Elizabeth to cut my hair. She reluctantly agreed, and I'm happy to report that I look nothing like a chicken, plucked or otherwise. I considered growing a bushy mustache, but vanity got the best of me. Albert Einstein, Tom Selleck, Dr. J., and porn star Ron Jeremy—they had some of the most iconic mustaches of the twentieth century. That kind of facial hair takes a confidence I just don't have. So, I looked to other ways in which I could embody an Einsteinian lifestyle.

I stumbled across a website for a German typographer, Harald Geisler, who spent six years developing a font based on the handwriting of Albert Einstein. "I like to imagine that when one uses Einstein's handwriting as a font that a spark of his genius potentially could reflect in one's own writing," he said. I downloaded it and started using it for everything I wrote. My grocery lists were now modeled on Einstein's scribblings, something that ended up annoying my wife. Maybe that explains Einstein's multiple marriages.

Geisler had contacted the Einstein Archives to see as much of the physicist's handwriting as possible and was so meticulous in his design that he created five variations of each letter—five As, five Bs, five capital As, five capital Bs, and so on. "My idea was to have different versions of every character and when you type, these characters change, so you never have the same characters next to each other," he explained. "If you write 'look,' for example, it has two Os, but the Os will be different." Geisler was voracious and kept coming back to the archives for more. "He said he needed more Qs," curator Roni Grosz told me with a laugh. And Geisler is not done. He is still working on improving the Einstein font. In addition to the five variations of each letter, he's now designing a set of Greek symbols and numbers based on the mathematical equations written by Einstein.

I'm finding habits I have in common with Einstein. We both despise beer. Einstein never skipped breakfast, which is terrific, because that's my favorite meal of the day. So much so that I often will have breakfast for dinner. He reportedly had mushrooms and two fried eggs on most mornings—food that was nutritionally dense and high in antioxidants. In other words, good brain food. We raise a flock of chickens whom we endearingly refer to as the Co-Hens and have bestowed upon them names of our favorite NPR broadcasters. Terry Gross laid our very first egg and Yuki Noguchi is the friendliest. Kai Ryssdal, our rooster, can be a jerk. To make my menu more authentic, I got my breakfast eggs from a hen we call Alberta Einstein. Google "Polish chicken" and you'll get the picture. She's completely black except for a huge pompadour of wild white hair atop her little head. Starting each day without having to consider what to wear or think about what to eat, was already improving my productivity.

But not all of Einstein's eating habits were healthy. Like me, Einstein never met a baked good he didn't like. That may explain why he filed paperwork with the U.S. Patent Office for an expandable shirt. (Seriously.) And he was just as absent-minded while snacking as he was with every other part of his life. Einstein's book club friends once put out bowls of caviar, which he quickly devoured. "Do you know what you've been eating? For goodness' sake, that was the famous caviar," they said. Einstein's reply? "Well, if you offer gourmet food to peasants like me, you know they won't appreciate it."

Einstein came up with many of his best ideas by devising thought experiments and would often daydream to help figure out complicated concepts. He believed that you could stimulate new ideas by allowing your imagination to wander, without worry about existing constrictions. When you set out to solve a problem you are, in effect, within the confines of that question. But Einstein gave himself permission to just let his mind wander. He often played classical music on his violin as a brainstorming technique. There is science to back up this method. Researchers at the University of California in Santa Barbara studied a group of physicists, artists, and writers and asked them the following: When that "aha" moment came, what were you doing? Turns out that many of them had their big discovery when they were focused on something mundane.

It's one of the reasons Einstein loved to go sailing. He found peace and serenity aboard boats, away from the adoring crowds.

"A cruise in the sea is an excellent opportunity for maximum calm and reflection on ideas from a different perspective," the physicist wrote while floating near the shores of Panama. His wife, Elsa, added: "There is no other place where my husband is so relaxed, sweet, serene, and detached from routine distractions, the ship carries him far away." And so, I've decided to write this chapter while on a boat in the middle of a lake in my town. Surrounded by nature and the occasional jet-skier, and with poor cell service and no internet to distract me, I'm beginning to see what Einstein was talking about.

By all accounts, I'm a better sailor than Albert Einstein. Which isn't saying much. Sailing was one of Einstein's favorite pastimes, yet the irony was he wasn't really that good at it. "According to his biographers," wrote Philip R. Devlin, "he would lose his direction, his mast would often fall down, and he frequently ran aground and had near-collisions with other vessels." What's more, Einstein rarely wore a life vest and could not even swim. This meant he was constantly having to be rescued— sometimes by kids and other times by nearby boaters. There's a classic headline from the *New York Times* that reads: "Relative Tide and Sand Bars Trap Einstein." Another headline read: "At relativity, a genius; as a sailor, not so much."

But for Einstein, who called sailing "the sport that demands the least energy," it was an escape. "He used to sail as a way to think. He loved the serenity of it," said Jim Lynch, whose novel *Before the Wind* opens with this line: "Einstein wasn't a great sailor, probably not even a mediocre one." As Walter Isaacson noted in his biography of Einstein, "He usually went out on his own, aimlessly and often carelessly. 'Frequently he would go all day long, just drifting around,' remembered a member of the local yacht club who went to retrieve him on more than one occasion. 'He apparently was just out there meditating.'"

Lynch told me that there are lots of stories of Einstein being towed back in. "I found ample amusement in the fact that the genius of physics is a little bit confounded by the physics of sailing. His doctor, at one point, said he didn't think sailing was good for his heart, but Einstein kept sailing." Albert wasn't the only Einstein who had a wobbly seaside reputation. Lynch recalled hearing a story about Hans Einstein, Albert's oldest son. "Everybody felt it was too dangerous to go sailing in the bay with Hans. It was like a family love for a sport that they didn't do well. Which is kind of endearing."

When Einstein turned fifty, in 1929, the city of Berlin honored its most famous citizen with a lake house. "It's the ideal residence for a person of creative intellect and a man fond of sailing," a relative noted. Einstein agreed, writing to his sister that, "The sailboat, the sweeping view, the solitary fall walks, the relative quiet—it is paradise." As Lynch writes of Einstein in his book: "He didn't race or cruise, but he understood the pleasing mix of action and inaction and the thrill of a sunset into the spangled bliss."

When Einstein would be bobbing around the Long Island Sound, getting lost both geographically and in his thoughts, he would let his mind wander so that he could be uninhibited by established norms. Those "aha" moments can surface when you let your subconscious roam without judgment. "Imagination is more important than knowledge," Einstein once said. He looked at the world—whether it was clothes or the cosmos—from an entirely different perspective than the average person. He tuned out conventional wisdom and questioned everything. "Einstein's secret was that he asked really good questions," explained the *New York Times* cosmic affairs correspondent, Dennis Overbye, who wrote a book about Einstein. As Alex Trebek, the longtime host of *Jeopardy!* said: "Answers can kill. But questions can take you anywhere."

In his book *The Puzzler*, AJ Jacobs calls this the "puzzle mindset" and suggests a mantra we all follow: always be curious. "If I hear about the climate crisis, I want to curl up in a fetal position in the corner," Jacobs wrote. "But if I'm asked about the climate puzzle, I want to try and solve it." A team of scientists using eye-tracking software discovered that creative people were able to physically see more through their eyes. It's about training your perceptions to bring in new information. The more rigid your worldview, the more you put outside of your scope.

Adam Grant, noted author and psychologist at the Wharton Business School at the University of Pennsylvania, says there are four types of thinkers: preacher, prosecutor, politician, and scientist. Only when you think like the latter do you "favor humility over pride and curiosity over conviction," Grant explained. "You look for reasons why you might be wrong, not just reasons why you must be right." Einstein viewed his opinions as a series of hypotheses to be proven right or wrong; the final answer is, in some ways, less intriguing than the search for answers itself.

Thinking like a scientist means being able to change your

mind. You know what you don't know, and you're eager to discover new things. It's the exact opposite of a preacher who is convinced he's right, or a prosecutor who is trying to convince you of a case, or a politician who is just trying to win your approval. Einstein had humility in spades and often was self-deprecating or mocked those who elevated his status. "As for the words of warm praise addressed to me," Einstein said, "I shall carefully refrain from disputing them."

We all suffer from what Grant calls "cognitive entrenchment," when we think we're so knowledgeable about a particular subject that we fail to ingest new information because we're so stuck in our ways. What's more, being wrong can actually lead to a happier life. "If you can embrace the joy of being wrong," Grant said, "then you get to anchor your identity more in being someone who's eager to discover new things, than someone who already knows everything."

Albert Einstein received one of his most prized possessions when he was just five years old. It was from his father, and it was a compass. Young Einstein so desperately wanted to know why its needle pointed north. He was fascinated by the invisible force behind it. "Curiosity, in Einstein's case, came not just from a desire to question the mysterious," wrote Isaacson. "More important, it came from a childlike sense of marvel that propelled him to question the familiar." I'm trying to incorporate this mindset into my everyday life.

One of the things that made Einstein transcend the science community is that he was so much more than just a physicist.

He had myriad interests: In addition to sailing, he loved to bicycle. (That could be because he never bothered to get a driver's license.) He was offered a job as an insurance salesman. He handed out candy to local children and helped them with their homework. In his spare time, he invented a noiseless refrigerator.

Nothing seemed to slow him down. He spent a decade working on crystallizing his theory of relativity. I can't even remember if there was ever a time I stuck with a problem for that long. There is so much we can learn from that mental stamina.

In 1946, and after completing his studies in chemistry at UC Berkeley and serving as a navigator in the U.S. Air Force, Clymer Marlay Noble Jr. was unsure what to do with his life. So, he sent a letter to Albert Einstein asking for career advice. "The main thing is this," Einstein wrote back. "If you have come across a question that interested you deeply, stick to it for years and do never try to content yourself with the solution of superficial problems promising relatively easy success." Noble decided to pursue a career in chemistry and worked in the field for thirty years.

And another trait of Einstein's that I aspire to: he was an indefatigable friend. David Bohm, a fellow physicist, left his post at Princeton and moved to Sao Paulo, Brazil, for a teaching position. The only problem? He was miserable in South America and almost immediately regretted his decision. He wrote letters to Einstein seeking advice and counsel, both of which Einstein was willing to dole out. Here are four things he said to cheer up his sad friend.

1. Be happy: "Although I fully understand your feeling of frustration," Einstein wrote, "I feel that patience, combined with an attempt to enjoy your life there as well as possible, seems to me the best you can do for the moment."

2. Hire a cook: Bohm complained about many parts of life in Brazil, including the local cuisine. Einstein empathized with his friend. "What impressed me most was the instability of your belly, a matter where I have myself extended experience," Einstein wrote, before suggesting that Bohm hire a "reliable cook."

3. Stay put: Bohm asked Einstein if he should bail on his Brazilian adventure and seek employment in another country. He wrote back the pros and cons of many places, saying "Israel is intellectually alive and interesting" and that, whatever you do, don't move to Ireland. Einstein added that "in spite of this, you should, in my opinion, hold out until you have acquired Brazilian citizenship." (Bohm would eventually move to Israel, where he met his wife.)

4. Put things in perspective: Between worrying about meals and moving, Bohm was also desperately trying to solve a scientific equation but was feeling discouraged. So, Einstein wrote back: "You should not be depressed by the enormity of the problem. If God has created the world, his primary worry was certainly not to make its understanding easy for us."

Such thought and effort went into this correspondence. Meanwhile, I can barely muster the energy to send a short text back to my friends. It's no wonder, then, that Einstein's letters to Bohm sold at auction for $210,000.

Back out on the lake, I have, unlike Einstein, not gotten turned around on this boat. The fresh air, free from distraction, has been mind-expanding. And I am returning to shore—without the help of nearby children—with more knowledge than when I left.

WITH A NAME LIKE EINSTEIN

"Many things which go under my name
are badly translated from the German
or are invented by other people."

—ALBERT EINSTEIN

Having the last name Einstein is a little like being Bob Galileo—an actual person, by the way, who's a production manager at a 3D-printing facility and lives in Aurora, Illinois, according to his LinkedIn profile. It's likely the first thing people ask him at a party: Oh, are you related to *the* Galileo? It's the ultimate icebreaker.

But Bob Galileo has nothing on Harry Einstein, whose rise to fame coincided with that of his distant cousin. When Albert was out winning a Nobel Prize, Harry was receiving accolades of another kind. He began his career in the 1920s as the advertising manager at a Boston furniture store and performed stand-up routines for friends at parties after work. Eventually, they convinced him to give comedy a try, and he made a guest appearance on Eddie Cantor's popular radio program. Harry was a master of dialects and, for a skit on the show, he invented the Greek character Nick Parkyakarkus (pronounced *Park-your-carcass*). The audience loved him so much that he became a regular part of the show. That led to guest appearances on Al Jolson's show and eventually headlining his own radio show for NBC, *Meet Me at Parky's*.

On the night of November 23, 1958, Harry was one of the featured performers at a Friars Club roast of Lucille Ball and Desi Arnaz. More than one thousand people were in attendance at the Beverly Hilton hotel in Los Angeles. The evening's emcee, Art Linkletter, introduced Harry and he took to the podium. Harry's set was only scheduled to be six minutes long, but the constant laughter from the audience drew it out. It lasted ten minutes, and it was quite literally the performance of his lifetime. "It was the most hilarious routine I ever heard," Milton Berle later said.

Harry wrapped up and walked back to his seat on the dais. And then he promptly keeled over onto Berle's lap. When Berle shouted from the stage, "Is there a doctor in the house?" the audience initially thought it was part of the act.

The doctors in the audience—there were many as the event doubled as a charity benefit for local hospitals—dragged Harry's body backstage to examine him. One reportedly took a pocket knife and opened up Harry's chest to massage his heart. Meanwhile, with the audience unsettled, Berle asked Tony Martin, a popular singer, to calm the audience down. Martin, ad-libbing and under pressure, chose an unfortunate song: "There's No Tomorrow."

Harry Einstein died that night at fifty-four.

Seeing your father die on stage while everybody around you laughs is sure to leave an impact. Perhaps in spite of, or because of, those twisted circumstances, two of his children went into show business, becoming comedic actors—although both changed their names. Bob Einstein started going by Super Dave Osbourne, the Wile E. Coyote of stuntmen, and appeared regularly on *The Smothers Brothers Comedy Hour* and other variety shows. Younger audiences may remember him for his final role, before his death in 2019, as Marty Funkhouser on HBO's *Curb Your Enthusiasm*.

Bob's younger brother felt a more compelling reason for a stage name—after all, who would take an actor seriously if his name was literally Albert Einstein? So, he became Albert Brooks. Yes, the Academy Award–nominated star of *Broadcast News*, among a myriad other movies, was actually born Albert Einstein. Brooks likes to joke that "the real Albert Einstein changed his name to sound more intelligent."

But not everyone in the family felt the need to shy away from the Einstein name. Bob and Albert had a half brother (from Harry's first marriage), Charles Einstein.

Charles was a peripatetic and prolific sportswriter, working for the International News Service. Back then, in the late 1940s, reporters in the field would dial the company switchboard, the operators would connect them to someone on the news desk, and the reporters would dictate their stories over the phone. One evening, Charles was working the sports desk while his mentor, a reporter named Paul Allerup, was out covering an event. After the game, Allerup phoned the INS switchboard. "Get me Einstein," he told the operator. The operator that night had never heard of Charles Einstein, so she checked her famous name rolodex and patched Allerup through to Princeton, where Albert Einstein picked up the phone.

"Einstein?" said Allerup.

"Ya," Einstein replied.

"Take this down," Allerup barked, then proceeded to dictate his entire story to Albert Einstein, who never once revealed his complete identity. The next morning, as was his habit, Allerup stopped by the newsstand on his way to the office to see which English language dailies picked up his wire story from the night before. Needless to say, his story was nowhere to be found. Frustrated and loaded for bear, Allerup arrived at the office and headed straight to Charles's desk. "You son of a bitch!" he yelled. "You were drunk last night!"

"I wasn't drunk," Charles protested, completely taken aback.

"The hell you weren't," Allerup told him. "And what the hell was going on with that crappy German accent?"

Later that day, Allerup was still trying to find out what happened to his wire story when he confronted the phone operator who patched him through by mistake to the world's greatest scientist. "You put me through to the wrong Einstein," he said to her.

She simply shrugged. "Any sane person," she said calmly, "probably wouldn't complain."

Charles moved his family around a lot, depending on where the sports reporting took him. They went from New York to Scottsdale and eventually to San Francisco. "We have a history of building and buying places, and then selling them for pennies on the dollar that later on turned out to be million-dollar homes," Charles's son, Jeff, told me. "Part of the Einstein business savvy."

In California, Charles became close friends with future Hall of Famer Willie Mays and eventually authored a biography about him, called *Willie's Time*, which was a finalist for the Pulitzer Prize. Charles would bring Jeff up to the press box at Candlestick Park and introduce him as the next senator from the state of California. By the time Jeff was in middle school, and the young boy was not showing the leadership skills of a politician, Charles pivoted and said Jeff would be the next great lawyer from San Francisco. "By the time I graduated high school, my dad would turn to people and say, 'I want to introduce you to my son, Jeff. At least he's not in prison.' So, my whole childhood was an exercise in diminished expectations." (The family pet, a Dalmatian named Albert Einstein, had better prospects. "He used to get more mail than anybody in the family and eventually wound up with the best credit," Jeff said with a laugh.)

Indeed, Jeff did not end up in prison. But he did end up on the run. Like Albert Einstein (the human, not the dog), Jeff was a pacifist. After high school, to avoid being drafted into the Vietnam War, Jeff fled to Israel, which accepted him with open arms under the Jewish Law of Return. But that didn't last long. Several months after the Yom Kippur War of October 1972, the Israelis tried to draft him into its army. "I did the adult thing," Jeff recalled, "and caught the next flight to Frankfurt. So, I've been chased out of two different countries by two different armies."

He went to work copywriting for ad agencies. One day, he grabbed a book near his desk—a dense word processor manual—and brought it to the bathroom. "I have a theory that the reason why there are many more men grand champions in *Jeopardy!* than women is because men will read anything on the throne," he said. But he couldn't understand a word of it, and that's when he had an epiphany. Despite knowing nothing about how computers work, he thought he could write a better manual than the one he was holding while sitting on the toilet. When he got back to his desk, he wrote up a proposal for an entire series of *Einstein's Beginner's Guide* computer manuals and sent it to various publishing houses. A few weeks later, he got a call from the acquisitions editor at Harcourt. "Mr. Einstein, this is a pretty unusual proposal. It's not too often we get someone pitching a seven-book series based on their ignorance."

Harcourt asked if his last name was really Einstein. "It used to be Hemingway," Jeff told them. "But no one reads Hemingway anymore."

The books came out and were a success. It was the mid-1980s—when personal computing was beginning to take shape—and people were desperate to learn how to use this new device in their home. Jeff went on media tours and got endorsement deals. Along with a friend, he created one of the country's first digital ad agencies. They would make corporate cartoons, then send them out in the mail on floppy disks to thousands of businesses and consumers. Jeff eventually tired of the grind and moved to Hawaii for a few years. When he returned to the East Coast, he enrolled in cooking school. "I figured knives and fire would slow me down a little bit," he said.

By now, the 1990s had arrived and the World Wide Web was coming online. Jeff's expertise, as an early crusader for digital advertising, made him a shaman in Silicon Valley. The one-time computer novice was now in high demand. "They paid me God's money," Jeff said. The *New York Times* called him "the Mick Jagger of digital media."

But it all came crashing down after the dot-com bust. He was laid off in May 2001 and ended up selling khakis at the Gap for ten dollars an hour. In the decades since, he's rebounded: selling real estate in Manhattan, and selling cars in Florida, where he is now semi-retired.

All of the job changes caused stress on his marriage. His wife, Mara, wanted a divorce. In the settlement negotiations there was, however, one thing she wanted to keep: the last name Einstein. She had grown up on Long Island with the "Smith" of Jewish last names: Schwartz. "A choice between Schwartz and Einstein really wasn't a fair fight," she told me.

Mara is a published author and media studies professor at Queens College in New York. Her students call her Dr. Einstein. Changing her name was out of the question. She loved being known as Einstein so much that when she remarried, she kept her ex-husband's last name. At conferences, when she gives talks, she always opens with the same line: "I'm Mara Einstein, and I'm going to answer the question I know you are all thinking: No, I'm not related, but, yes, my daughter is."

While it doesn't help her score dinner reservations—"If it was Kardashian, maybe,"—she said it does help people remember her in academic circles. (Jeff told me he once made a reservation at a Manhattan eatery for himself and a friend whose last name was Shakespeare.)

Before entering academia, Mara worked in television marketing. She woke up on Friday mornings and the first thing she did was check the ratings for the previous night's episodes of *Friends, Seinfeld,* and *ER.* After enough Fridays had come and gone, she came to realize the mundanity of what she was doing with her life. "Getting people to watch more television couldn't be the best use of my energy on the planet," she said. So, she took a page out of the Einstein playbook and decided to become a professor. Like her ex-husband, she was inspired by a book in her bathroom: this one was a collection of letters that children had sent to Albert Einstein. "He had a very grandfatherly reputation," she said. "Somewhere, there developed a sense of him not just as a theoretical physicist, but as somebody who was willing to be a teacher."

She focused her research on how media, marketing, and celebrity culture impact society. In talking about Albert Einstein's fame, she explained that "people may know $E=mc^2$, but they have no idea of what it means. That idea that his thinking was beyond something that they could comprehend, makes people look up to him. Yet at the same time, I think there's a sense of that very neat, very tidy equation, where he somehow took something very complex and made it quite simple for a lot of people. It makes him feel approachable."

In essence, she explained, successful advertising does the same thing: it takes complex ideas and presents them in an accessible way. General Electric, for example, may manufacture complex machinery that produces renewable energy, but the way it advertises that to the average consumer is through showing beautiful vistas with wind turbines. "If you had to go to central casting and pick the quintessential look of a scientist, Einstein would be it. You would pick somebody who's not particularly neatly dressed—not a schlub, but not totally put together—and has hair that looks like he stuck his finger into a light socket."

Add to that, timing. Einstein came to prominence not only during the golden age of radio, but also as television sets were becoming more common in people's homes. In the 1940s, about seven thousand homes had a TV set. By the 1950s, that number was close to seven million. And the people on that TV—Milton Berle, Dick Clark, and, yes, even Albert Einstein—were the biggest stars of the day. "This sense of celebrity and wanting to be able to reach out and touch a movie star at that time was such a big deal," Mara said. "So, imagine seeing someone on the street who's really famous, that you could reach out and touch. The level of interest and awe and fandom that that could engender is probably beyond our ability to even comprehend at this point." Her students, who never lived in an age without cell phones or social media, can't imagine a time when an entire society consumed all of the same pop culture. Bumping into Einstein walking on a Princeton sidewalk in his pajamas was a tangible experience that could happen. The same can't be said for Beyoncé.

That next generation includes Cayla Einstein, Jeff and Mara's only child, and the great-great-grandniece of Albert Einstein.

Born in 1999, she said growing up with the last name Einstein was difficult. It didn't help that the one class she ever failed in high school was physics. She had a poster on the wall of her bedroom—that famous photo of Albert Einstein sticking out his tongue—as if the genius himself was mocking her. She got made fun of; there were some who outright asked her: "What's it like knowing that you'll never be able to live up to your name?" She attended the Bronx High School of Science, known for being a feeder to Ivy League universities, which didn't help. "I had an inferiority complex for a really long time," she told me. "Growing up, I had such a need to prove myself as this science person. I always went over the top in the science fair. I built a bionic leg." She waited a beat, and then added: "No one needed to do that. I was really setting myself up for failure."

That's not to say she didn't use her famous last name to her advantage. It was the topic of her college entrance essay where she wrote, "I am an Einstein, just not that Einstein." She admitted she wasn't smart when it came to science and math. When we were talking, she said, "I'm academically inclined, but there's just some things I have to acknowledge that I'm not going to be good at. And I'm not going to be Albert Einstein. For a lot of years, I was just like, 'What's the point?' because I'm just always going to be the subpar Einstein."

When she arrived at the University of Pennsylvania, she didn't even use her last name as an icebreaker when meeting new people. "It feels pretentious to lead with that," she said. "Everyone's going to think I'm just a crazy person." She had to take a required class in physics and the professor, adding insult to ignorance, mispronounced her name (Ein-STEEN). As if she didn't have one of the most prominent names of the twentieth century. On campus, she found her niche—embracing the artistic side of her personality. "Finding something that I'm actually good at and enjoy doing. And realizing I can be a different form of Einstein. I wasn't in this little box that I had to do quantum mechanics or whatever."

She now says that, when she gets married, she plans on keeping her last name. "It's kind of this funny thing about me. It's no longer an inferiority thing. It's an ironic thing."

According to the website HowManyofMe.com, there are fewer than one thousand people currently living in the United States with the last name Einstein. (One person, intriguingly, has the first name Einstein.) During my quest to find all the myriad ways Albert Einstein is alive and well in our current era, I've encountered at least a dozen people with the name Einstein—including not one, but two Rabbi Einsteins.

"I embrace it," Rabbi Stephen Einstein told me when I reached him in California. "Not a day goes by that this doesn't come up."

The retired rabbi, who served his Orange County congregation for thirty-six years, feels that the name has been a gift that he's glad to be able to pass on to his children and grandchildren. "Albert Einstein was not only recognized for his brilliance and what he accomplished in scientific fields, but I take even more pleasure in the kind of humanitarian he was and how he cared for people. I love being associated with that."

The rabbi, not surprisingly, has found Biblical lessons in his name. He recounts the verse in Deuteronomy 5:9–10, when God tells the Israelites that He will visit "the guilt of the parents upon the children, upon the third and upon the fourth generations of those who reject Me." But that God will "show kindness to the thousandth generation."

The rabbi waved his arms with excitement. "Wait a second, there's a math problem here. Right? And the math problem is, what if my grandfather was a horse thief? But what if my great-grandmother was a very righteous woman? Do I get the blame, or do I get the credit?"

He paused, briefly, to let the question sink in. "Of course, the answer is the prophetic word—*if*. If you carry on in the same way that your positive ancestor did, you're going to get credit. On the other hand, if you're acting in a bad way, then they're going to throw the book at you. So, it still comes down to individual responsibility."

As if to make his point, he added: "If Hitler happened to be your last name, you probably would change it."

Stephen's daughter, Rebecca, followed her father into the rabbinate. For a while, she worked at his congregation, making them one of the only father–daughter rabbinic duos on record. But when I caught up with her, she was living in the Lehigh Valley region of Pennsylvania, not a hotspot for Jews or their clergy. "My children are considered ethnically diverse because they don't eat bacon," she said.

Rebecca said that Einstein was a "super weird last name to grow up with," and that it's strange to see the name on the signs of coffee shops (Einstein Bros. Bagels) and hospitals (the Einstein Medical Center in Philadelphia). People are always gifting her Albert Einstein T-shirts and magnets. "It's this mythos that surrounds our family."

But she's moved past all of that and, like her father, looks for inspiration in the man, not the name. Her son is autistic, and perhaps it runs in the family. "There's been a lot of armchair diagnosing, all these many years later, that Albert Einstein was probably on the spectrum. Perhaps he had Asperger's," she said.

When she speaks from the pulpit or leads spiritual discussions, she often brings up Einstein as an example of a modern man who is both a rational thinker and one who believes in a higher power. "He did have a lot to say about the coexistence of science and faith. That perspective has always fascinated me," she said. "It's a good reminder that there are all of these dualities that exist both in creation and within ourselves."

Rebecca sees in Einstein something deeper. "His understanding of energy and the notion that energy can neither be created nor destroyed, but simply transferred. And that, to me, has always been a very powerful metaphor for a person's soul," she said. "After our body, which is a divine creation, is no longer needed, we put the body to rest in the Earth, but the soul goes back to the creator. It helps people reconsider what they thought a soul is and how it's different from the idea that one moment you're here and then you're gone."

She continued: "I love that with all of Einstein's knowledge, he still experienced that awe and wonder that even with all of his understanding of the universe, he still understood that he didn't understand everything, and that there was something more and greater and powerful. And he called that 'God.'"

If you're trying to track down all the Einsteins living in America —which is what I'm trying to do at this very moment—the best place to start is with Laurie Einstein Koszuta. She's one of the world's foremost compilers of Einstein genealogy. Ask her about any Einstein and she'll tell you how they're related. When we talked, she had 5,346 living and deceased Einsteins on her family tree, enough for a forest.

Laurie's obsession with her last name began in high school. This was in the 1970s, in Indiana, long before the days of Ancestry.com. While her friends were hanging out at McDonald's after school, she would take the bus to the local library to do research. She discovered the first-known Einstein—Baruch Moses Einstein, born in Germany in 1665. He was the great-great-great-great-great-great-great-grandfather of Albert. As luck would have it, Fort Wayne's downtown library had phone books from other cities, and she would look up people named Einstein, take down their names, and write to them. "Back in my day," she said, "calling long distance was expensive."

To her surprise, she received letters back, a tradition that continues to this day. During the height of the COVID-19 pandemic, when everyone was stuck at home looking for things to do, she received emails from people named Einstein. She told me that she's not even the only Einstein where she lives in Florida: she attended the same synagogue as George Einstein, an engineer and accomplished cellist who occasionally played in a band with Albert Einstein. (George died at ninety-one in 2011.)

She said that for people who don't know her, the name can sometimes be off-putting. "I think on Facebook, people say, 'Oh, she thinks she's so smart because she sticks Einstein in there.' I don't think they realize it's my maiden name. It's like putting 'The Dog' or whatever nickname you decided to stick in there. And I have to tell people, 'No, that's really my name." Even her friends rib her. "They'll say 'She's an Einstein, she's smart, she can figure that out.' And I'm like, really? I have to use an accountant also, believe it or not."

She told me about Arik Einstein, the most influential Israeli singer of the modern era. His death, in 2013, was covered by the media as breaking news, reminiscent of Michael Jackson being rushed to the hospital just a few years earlier. She pointed me to Daniel Einstein, a historic preservationist at the University of Wisconsin-Madison. Daniel had been working on tracing his family tree since he was in elementary school when his friends started joking with him about his name. "I used to call it my theory of relatives," Daniel said when I reached him. Albert Einstein was his grandfather's fourth cousin. "Albert dipped deep into the gene pool and grabbed all the good stuff and really didn't leave anything for us mortals." And yet, there was still enough smarts to go around. Daniel's great-grandmother was the cousin of another Nobel Prize winner, Joseph Erlanger.

When Daniel arrived on campus in 1990, one of the cafeterias was called Einstein's. It sold an atomic burger; the menu listing used a caricature of Albert with a mushroom cloud over his head. Needless to say, Daniel took umbrage that they were trivializing his cousin's likeness to make money. "First of all, Einstein was a pacifist," he told them. "And the association with the nuclear bomb is a historic travesty. What other historic character might you choose to exploit? Would you have a Martin Luther King hot fudge sundae?"

It was the historian in him coming out. And he's not saying that Einstein doesn't deserve some sense of scrutiny. "History is a story that can always be made more authentic, with more perspectives and more opportunities to get viewpoints from a variety of sources," he said. "But Einstein is always held up to be this iconic, perfect person. And then you learn a little bit about his extramarital relationships and his mistreatment of his first wife. And you say, 'He was imperfect, but that that makes him human.' It doesn't make his behavior admirable by any means, but it does provide a fuller understanding that everybody is multi-dimensional, literally."

The game of Einstein phone tag continued when Daniel, in turn, told me to call his cousin Ted at the University of Maryland, who is an actual physicist with the last name Einstein. "It's a double-edged sword," Ted said. "For me, one of the criteria for getting tenure was you need a national reputation. And I think it was somewhat easier for me because people tended to remember my name. But then it cuts the other way, that chuckle and snort, people don't take you as seriously. It's sort of embarrassing in some ways, too. Nothing is good enough compared to what Albert did."

As you can tell, many Einsteins do, in fact live up to their name, becoming teachers in institutions of higher education. Sarah Einstein is an English professor at the University of Tennessee at Chattanooga. Her Twitter handle is @SarahEMC2. "My students are always really disappointed that their Dr. Einstein teaches creative writing, not physics," she said.

However, she is one of the few Einsteins I met that has a flawed view of the name. "I come from the asshole side of the Einstein family," she told me, rolling her eyes. She grew up in Appalachia, in the foothills of West Virginia. Her family, Jewish peddlers from Lithuania by way of Baltimore, ended up there by chance. One of her ancestor's carts broke down. "He built a general store because it was cheaper than fixing his cart."

That was on her mother's side. "My father," the one named Einstein, "was raised Presbyterian. My father's father was a virulent racist and antisemite. I can remember him introducing me to his golf buddies as his little kike granddaughter." Her great-grandfather, John Einstein II, was just two years younger than Albert. During the First World War, John had a crisis of his Jewish faith and enrolled in divinity school where he was ordained a Presbyterian minister. But, according to Sarah, he was a far cry from a man of God.

Like the traveling salesman in *The Music Man*, John flitted around from town to town. "He denomination hopped," Sarah explained. At one point, he was a pastor in Virginia where he raised a lot of funds to build a church and then took off with the money. "He was so divisive, that they almost dissolved the

Unitarian Church in Richmond in order to get rid of him. He was just an asshole, as were many of the Einstein men in my family. My Einstein grandfather was a really abusive son of bitch. I'll be honest, they were pretty terrible. They were grifters." She calls them the "poisoned branch of the family tree."

Appalachian. Antisemites. Assholes. I've got to admit: Those three words were not on my Einstein Bingo card.

She's now trying to come to terms with her last name and, in 2022, traveled to Ulm, the German town where Albert Einstein was born. She learned about her thrice-great-grandfather, Moses Leopold Einstein, and saw photos of other relatives like Arthur Einstein and Irene Guggenheimer Einstein. These people, she described, were the "good" part of her Einstein family tree. "I *do* feel kinship with the people in these pictures, but I question it," she wrote in a diary on the trip. "Am I allowed fellow-feeling with the Einstein family as a larger whole if I feel none at all for my father's parents or his sister and her family? Or, by virtue of how awful they were—and how much harm they would have wished on these other Einsteins, because they were virulent antisemites—is it wrong, perhaps selfish or self-aggrandizing, to want to also know this part of my family history and to call it that: my family history?"

Sarah has been married multiple times and has had ample opportunity to take on a new husband's name. But she still keeps her maiden name—from the aforementioned asshole side of the family. "I'm just enough of a feminist that I feel like it's kind of weird to give up your name and your identity." In addition, none of her husbands have been Jewish. "It was more about holding onto my Jewish identity than it was onto the Einstein part," she said. "And I felt like it would erase it to have a non-Jewish last name."

I ask the inevitable question I ask all the Einsteins I interview: Are you related to Albert?

"Ish," she replied with a smile.

She is fourth cousins, three times removed, adding, "Which means that I'm not related. Even the Appalachian part of me is like that's pretty damn distant."

Colonel Max Einstein—a brigadier general who fought gallantly in America's Civil War, who was personally assigned to his post by Abraham Lincoln and received a hero's welcome in local newspapers—was probably a bit jealous of his distant cousin Albert who, you know, garnered just a tad more attention than he did. High school students in the twenty-first century are still studying $E=mc^2$. Max, on the other hand, well, I would venture to guess this is the first you're hearing of him.

Max was born in the same German town as Albert's father, emigrated to the United States in 1844, and quickly made a name for himself in his adopted city of Philadelphia. His ribbon and silk store in the downtown district of the City of Brotherly Love earned him wealth and political connections. Max was so close with Simon Cameron, Lincoln's Secretary of War, that Max named one of his sons after him. (According to a quick search online, a possible descendant named Simon Einstein is the proud owner of a surf shop in Portugal.)

Einstein became involved in local Pennsylvania militias, eventually earning the rank of lieutenant colonel working side by side with Pennsylvania's governor. By 1860, as the Civil War was mounting, Einstein was elected the brigadier general of the Second Brigade of the Pennsylvania militia. The following year, he organized a group of more than one thousand men to join the war effort—often spending his own fortune to help train and outfit the soldiers. According to historians, at least half of the regiment were German immigrants and felt a close kinship with Max. They sprang into action in the spring of 1861.

At the first battle of Bull Run at Manassas Junction in Virginia, Confederate General Stonewall Jackson defeated the Union Army. As the soldiers from the north retreated, Max Einstein stayed behind to make sure his compatriots could escape to safety. "Colonel Einstein...returned to the field of battle at eleven o'clock on Sunday night and brought off six pieces of artillery, which he delivered to the commanding officer on the Potomac yesterday evening," read the detailed report in the *Philadelphia Inquirer*, adding that Einstein also captured eighteen horses belonging to Confederate troops. "Colonel Einstein reports that the field was then clear, and not an enemy in sight."

His courage on the battlefield that night would be the pinnacle of his career. He was later jailed (it's a long story), and when Lincoln appointed him to be the U.S. Ambassador to Germany, the Senate roundly rejected the nomination. Max blamed antisemitism for his treatment. He died a pauper on April 1, 1906, and was buried at Mt. Sinai Cemetery in Philadelphia. His plot is in the section belonging to Congregation Keneseth Israel, a synagogue at which his better-known cousin Albert would later become an honorary member.

One particular University of Alabama souvenir shop is, by any estimation, in the strangest of places. That's because it is exactly 6,598 miles from the campus in Tuscaloosa. Hundreds of gift shops line the cobblestoned streets of the Old City section of Jerusalem, where tourists flock to purchase holy items: like prayer shawls, anointing oil, and a tiny tube of sand from a nearby archaeological dig. It is also where you can buy T-shirts and other knickknacks that display the University of Alabama logo in Hebrew. A sign that reads WELCOME TO BAMA COUNTRY greets customers. Hani Imam, a 1989 graduate of the school, opened the store in 1995 to honor his alma mater. It is believed to be the only shop in Jerusalem dedicated entirely to an American university. It is also where, one spring day, I bumped into Karen Cortell Reisman, Albert Einstein's cousin.

We had met the previous day at the Albert Einstein Archives on the campus of Hebrew University in Jerusalem. We were both in town to attend a week of events celebrating the 140th birthday of Einstein, which was capped off by a press conference—where Reisman spoke—and the unveiling of 110 new documents, including eighty-four sheets of equations scribbled by the world's favorite mathematician. We hugged and I ushered her out of the shop just as Imam was trying to hard-sell Reisman a scroll with Bible verses printed on it.

The afternoon heat could already be felt. "The only thing I inherited from Albert Einstein is his hair texture," she said, ruffling her fingers through her blond mane. "I yearn for days of low humidity."

Reisman might be one of the only Einstein cousins I've met who has made a career out of being related to the Nobel Prize physicist. And she's done it in the most creative way. The Dallas-based Reisman—confident, gregarious, and a natural storyteller—is a motivational speaker. She runs workshops about life lessons people can learn from Albert. "I cannot explain the theory of relativity," she tells her audiences, "but I do have a relative's theory on Einstein's human side."

One of the first stories she always brings up is how Einstein and her grandmother, Lina, corresponded like "gossip buddies." You can see it in the notes themselves. "I don't use quite as many exclamation points when I write, but she had an exclamation point at the end of every sentence," Reisman said. One of her favorite possessions is a letter Einstein and her grandparents all wrote together on July 4, 1949, in Princeton. "Here they are, three European immigrants, celebrating Independence Day. And they're all happy to be in America." When Reisman got married and her mom asked her what she wanted as a gift, she had only one request: a copy of that letter. "It's one of my favorite things."

"His letters to my grandmother were filled with an update on their lives, an update on mutual friends, but he had a way of talking that was almost like a poet," she said. She rattled off, by heart, one particular line written by Einstein that has stuck with her: "About politics to be sure, I still get dutifully angry, but I do not bat with my wings anymore, I only ruffle my feathers." She said it's that line that teaches her a lesson about perspective. "When we are running around in our crazy, hectic lives, it's important to figure out when to bat with your wings, which means going full throttle, or when to just ruffle the feathers, when to sit back. Sometimes it's important to go full speed ahead. Sometimes it's wiser to sit back and observe, to think, strategize, and mull over what you need to be doing. This is a great metaphor on how to handle the craziness that's in your life."

When Reisman was growing up, she would tell her class-mates she was related to Einstein. One kid responded, "Oh yeah, my dad is Santa Claus." She didn't have much street cred. But as an adult, when she launched her company and started giving workshops to hundreds of business executives on how to better communicate and sell, she knew that using her Einstein connec-tion would set her apart. "Although I don't *just* use Einstein's sto-ries," she admitted. "That would be way too over the top."

As we walked amid the stalls in the Jerusalem marketplace, she gave me an example of a life lesson from Einstein that she teaches. "He had a wonderful way of finding the humor and being a bit self-deprecating about it by being humble," she said. "He knew he was a famous man. By the time I was born, he was an icon. But he wrote a wonderful letter to my grandmother when she was downsizing to a smaller place. He said, 'I had heard that you are about to move your point of gravity of your existence somewhat more into our vicinity.' I mean, that's hilarious! He's making fun of himself. To me, that shows his humor, and it's a reminder, it's a lesson to me to a) find the humor, and b) don't take myself so seriously."

Hanging out with Reisman is a joy, not the least of which because she is, in a way, a living simulacrum of Einstein. But she sees it as a simpler equation. "It just takes plain luck to have somebody famous on your family tree."

EINSTEIN AND POP CULTURE

"No living being deserves this sort of reception."

—ALBERT EINSTEIN TO HIS WIFE ELSA UPON
THE RECEPTION THEY RECEIVED IN JAPAN

The summer of 1989 was a good time to be a moviegoer. Tim Burton's *Batman* and the sequels to *Lethal Weapon*, *Ghostbusters*, and *Indiana Jones and the Last Crusade* had all just arrived in theaters. And let's not forget about classics like *Dead Poets Society*, *Field of Dreams*, and Spike Lee's *Do the Right Thing*. Then, of course, there was *Young Einstein*. The film starred a zany up-and-comer from Australia who had never been in a movie, and now here he was writing, producing, starring in, and directing his feature debut. The poster pictured the lanky thirty-six-year-old actor with a wild shock of orange hair that looked like the bride of Frankenstein had just stuck her finger into an electrical socket. His name was Yahoo Serious.

The ninety-five-minute comedy is a wide-eyed absurdist fantasy: Imagine if Albert Einstein grew up in Tasmania and was descended from a long line of apple farmers. In a drunken stupor one night, Einstein is trying to figure out how to add bubbles to beer. He somehow stumbles upon the solution: $E=mc^2$. "This will put the Einsteins on the map," his dad said proudly. "Albert, you're a genius."

The plot revolves around a rival stealing the famous equation to build the world's largest beer keg and who accidentally turns it into an atomic bomb. Along the way, Einstein invents surfing and rock and roll. (It has something to do with using the speed of light and a violin; a suspension of disbelief is obviously required.) Einstein falls in love with Marie Curie. The Wright brothers, Charles Darwin, and Thomas Edison all make cameos. Sigmund Freud, on-brand, shows up with his mother. In interviews, Serious described the film as "*Lawrence of Arabia* meets Bugs Bunny," and added that the "film is not a sendup of Einstein: it's a tribute."

Young Einstein took years to make and was filmed on such a low budget that Serious had to sell his car to help finance the movie. Unable to afford catering on set, his mom cooked for the crew. When it was released in Australia in December 1988, it became the country's sixth highest-grossing film of all time, besting *E.T.* and *Rambo*. Warner Bros. bought the rights to distribute the movie in the United States, hoping it would ride the popularity of that other Australian hit, *Crocodile Dundee*. Some American critics didn't quite get the humor of *Young Einstein*, with the *Washington Post* review calling it "dumber than a bowling ball." But others praised Serious for his physical slapstick, comparing him to a young Charlie Chaplin or Buster Keaton. It would become a cult classic. I personally remember renting a heavily used copy from our local Blockbuster. Serious, whose real name is the not-as-exciting Greg Gomez Pead, landed on the cover of *Time* and *Mad* magazines. He got his own recurring segment on MTV. He was even parodied on an episode of *The Simpsons*. Lisa Simpson saw a sign advertising the YAHOO SERIOUS FILM FESTIVAL and commented: "I know those words, but that sign makes no sense."

As it turns out, that would be the height of his Hollywood career. He followed up *Young Einstein* with two more movies— *Reckless Kelly* and *Mr. Accident*—and you could be forgiven for not remembering those. The latter holds the distinction of having only a 17 percent rating on Rotten Tomatoes, the review aggregator site. In 2000, perhaps in a bid to stay relevant, Serious tried suing the actual Yahoo! for copyright infringement, alleging that the popular search engine was profiting on his name. He cited, among other things, that 1989 *Time* magazine cover whose headline presciently proclaimed "Yahoo!" including the exclamation point. But the case was thrown out because he couldn't prove that he had suffered any financial harm from the tech giant. Perhaps he should've called himself Google Serious.

Serious pretty much vanished from public view since then. A reporter for *The Guardian* who spent decades searching for Serious said the actor's whereabouts was a bigger mystery than how the pyramids were built or who killed JFK.

Part of the problem was that Serious had eschewed the internet and had no online footprint. Which made tracking him down for an interview nearly impossible. I tried the usual routes: asking friends who work in Hollywood if they had any connections to him. They told me to look for Serious's agent or manager but, alas, he never had one—even at the peak of his fame. I've previously had success reaching out to celebrities via social media because messaging them as Albert Einstein catches their attention. It cuts through the chaff when, you know, a deceased Nobel Prize winner and someone more famous than them slides into their direct messages. (Before his death in 2023, I corresponded a few times with David Crosby of Crosby, Stills, and Nash.) But Yahoo Serious doesn't use Instagram. What made things more complicated is that, as far as I could gather, Serious lived half a world away as a recluse in a remote part of Australia. I called the one friend I know who lives in Melbourne to see if she could point me in the right direction, as if everyone in a country of twenty-five million knows each other. (Isn't Nicole Kidman your neighbor?) She laughed, as she took a bite from her vegemite sandwich—at least that's what I imagined she was doing.

Every few months—between interviewing a physicist looking for aliens and another physicist building a time machine—I'd spend some time going down a rabbit hole, researching online and off, trying to find Yahoo Serious. Two years went by and the deadline to finish writing this book was fast approaching. And that's when I finally had a break in the case. I stumbled across the website for the Kokoda Track Foundation, a humanitarian aid group that helps the indigenous people of Papua New Guinea, off the northern coast of Australia. Buried in the site was a page listing the non-profit's board of directors. And that's where he was hiding: There was a photo of Yahoo Serious in a helicopter, looking three decades older, but still with the shaggy hair and that distinctive smile. Is it possible Young Einstein was now a real-life superhero helping bring electricity to tiny villages?

I reached out to KTF and asked if they could put me in touch with Serious for a conversation over Zoom, but they told me he doesn't do interviews. Besides, they said, he's not that great with technology and doesn't like to be seen on camera. You'd think I was chasing down a story on par with Watergate or trying to photograph Bigfoot. But I had come this far and gotten so close; I wasn't about to give up. I kept contacting them until, eventually, they blinked and gave me Serious's email. (I can report it was not a Yahoo! address.) I sent him my request, and, to my pleasant surprise, he wrote back: "G'day, Benyamin." I nearly fell off my chair.

Yahoo Serious entered my Zoom room on a Wednesday in summer, but for him in Australia it was Thursday and winter. It was, in a way, poignant that our meeting about Einstein involved some sort of time travel. Serious's long flowing silver hair went past the collar of his thick gray cardigan, itself resting atop a powder blue V-neck sweater nestled over a slate blue shirt. His skin was ageless, his teeth a bright white. He was fit and trim. He looked like he stepped out of a J.Crew ad for seniors.

He was surprised when I told him how long I'd been searching for him. "I'm really not that hard to find," he said. Sure, if you happen to be one of the 1,593 people who live in the small beach town north of Sydney where he's camped out. He spends much of his time surfing.

He did make a rare public appearance in 2019 at a *Young*

Einstein screening in Sydney to celebrate the film's thirtieth anniversary. "It's like a good bottle of wine," he said of the cult classic. He brought to the event his fourteen-year-old daughter who had never seen the film on a movie screen. More than five hundred fans showed up. "It was like Beatlemania," Serious told me. "She was just so overwhelmed."

I asked Serious where the idea for *Young Einstein* came from and he dove into a fifteen-minute soliloquy venturing into all sorts of topics—Australian politics, discrimination against Aboriginals, the minds of dogs, the history of Greece, Shakespeare, Warhol, Che Guevara—and then would always return to the refrain: "Hmm, I forget where I was going with that." He admitted, on more than one occasion, that he's a lateral thinker. He kept losing his train of thought, very much like Einstein did. "It's a gift," Serious said with a sly smile. It's something that became evident when our scheduled thirty-minute chat meandered into two hours. By the time we were done, it was, for me, indeed Thursday.

"So, the idea for *Young Einstein*? How did you come up with it?"

"Right, right, man," he said in a mellifluous Australian accent, peppering his speech with the verbal tics of a full-time surfer.

Serious had been traveling with a buddy on a boat down the Amazon River in 1982. They were drinking a lot of beer when they came across a Brazilian man who, it just so happened, was wearing a ratty T-shirt that had on it a picture of Einstein sticking out his tongue. He said it all came together in that one instant, like a bolt of lightning. Serious took a photo with the man and still has it hanging on his wall.

Like most of us, Serious was aware of the broad strokes of Einstein's life, but there was one particular aspect that he connected with—Einstein's sense of humor and his penchant for making fun of himself. "We're Australians," Serious said, "so we make a joke of everything."

Upon returning home after that fateful trip, he did everything he could to research Einstein's biography. He read about the theory of relativity and the speed of light. He learned of Einstein's friendship with Charlie Chaplin, who was another hero of Serious. He was impressed with Einstein's prodigious "Miracle Year," when the up-and-coming physicist was just twenty-six years old. "When you've still got the mind of a child, but then start to get a lot of wisdom," Serious explained, "when those two things go bang together, it's great."

In many ways, Serious mimics his muse: Like Einstein, he has a childlike wonder for the world, is self-deprecating, and has a desire to run away from the spotlight (and yet somehow, they both ended up on the cover of *Time* magazine). They both made radical choices and stood up for what they believed in. They're both pacifists and, each in their own ways, humanitarians. Through his Kokoda Track Foundation, Serious is helping women in villages launch small businesses. He's installed solar panels so kids can have light to do homework at night. "Einstein was a reservoir of morality," Serious said of his inspiration.

He also tries to embody Einstein in another way, by having a passion for deep thought. It's why, he said, "I've never contributed to the internet in any way." He said he was recently watching a documentary that showed people's brains as they used technology. "Our brain is not designed to be annoyed all the time. So, as soon as you turn something on and—*ping!*—that's already destroying your brain. But Einstein was different. He would pull out his violin and think deep thoughts."

Serious, through his zigzagging discourse, mentioned three American presidents. "Those guys—Jefferson, Adams, and Washington—they all went back to their farms for winter." It's why, he said, he left Hollywood. "I turned down the money and turned down everything associated with that because I knew that I could understand the world better and speak as a global member of it better if I lived in my world that I was happy with, in my house overlooking my favorite surfing spot." He paused, and then added, "I don't think Einstein ever got out of bed saying, 'Oh, I'll do this because this will be really good for my career.'"

I asked him if he had any regrets. He said he would've liked to make a movie about another smart person of the early twentieth century, the psychoanalyst Carl Jung. He would've called it *Young Jung*.

Of course, *Young Einstein* is just one of the dozens of films that feature the world's favorite genius. My personal favorite is the 1994 romantic comedy *IQ* in which Walter Matthau portrays Einstein as a lovable uncle, quite literally. He plays cupid for his fictional niece, portrayed by America's sweetheart Meg Ryan, and an auto mechanic played by Tim Robbins. In a review of the film, Roger Ebert said Matthau deserved an Oscar nomination. Sadly, he did not get one.

Einstein is a powerful cultural myth in the world of Hollywood and in a galaxy far, far away. When Stuart Freeborn, a special effects supervisor on the *Star Wars* franchise, was looking for a model for the wizened face of Yoda, he said he based the character on Einstein. A time-traveling Einstein appeared in an episode of *Legends of Tomorrow*, a superhero series on the CW network. Science journalist Michio Kaku explains in his book, *The Future of Humanity*, that the warp drive used by the crew on *Star Trek* is based on Einstein's equations. But that's not the only time the physicist is referenced on that sci-fi series. In one episode, a main character is playing poker with three holograms: Isaac Newton, Stephen Hawking, and Albert Einstein.

Perhaps the most notable depiction of Einstein on TV occurred in 2017. The National Geographic channel aired a ten-part miniseries based on the life of Einstein. (Subsequent seasons of the *Genius* anthology series focused on other leading lights of the twentieth century: Pablo Picasso, Aretha Franklin, and Martin Luther King Jr.) At the time of its release, one of my tasks was to work with National Geographic in promoting the series on Einstein's social media pages. We would tease sneak peeks of upcoming episodes as well as behind-the-scenes videos of the making of the show. They sent me early copies of each episode so I could write a recap and have it ready to publish the moment the show ended.

Einstein was portrayed by Academy Award–winner Geoffrey Rush, who saw in the character a lesson for modern times. "I think the most obvious context around that is his passionate belief in the goodness of scientific inquiry," Rush told an interviewer. "That is something that we're battling with, in terms of major global debates about climate change, expenditure for research, or whatever, but I think he always had a rather golden quality, that science should only and can only be used for the goodness of human existence."

The series was made by the producing team of Ron Howard and Brian Grazer, known for such hits as *Apollo 13* and *A Beautiful Mind*. Together, they've been nominated for forty-four Oscars and more than two hundred Emmys. "I'd read other feature scripts that dealt with either Einstein's entire life or portions of his life, and they were never very effective," Howard said of his decision to make their version ten hours long. "You always felt robbed and cheated because it wasn't a kind of life that lent itself to a two-hour movie."

Released during the tumultuous years of the administration of Donald Trump, who was often antagonistic to the scientific research community, Gigi Pritzker, another producer, hoped the film resonated in ways it might not have before. "Having people like Einstein in the forefront of popular culture, and raising scientists to the level of celebrity, is a really important thing—so people who sit passively and don't think of themselves as scientists or understanding science can really get a grasp of why it's so important to support science and breakthroughs and research and development," Pritzker said.

For Grazer, the series held special resonance, given that he joined the project just as he was publishing a book about curiosity, which opened with a quote from Einstein. Grazer said he was influenced by Einstein in ways big and small. In the 1990s, as a way to stand out among other producers in Hollywood, he decided to grow wild-looking hair. "It lets people know that this guy isn't quite what he seems. He's a little unpredictable," Grazer explained. "I'm not a prepackaged, shrink-wrapped guy. I'm a little different."

One of the most popular police procedurals in Germany was a TV show about the fictional great-great-grandson of Einstein who helps detectives solve murder cases. The rights to the series have been sold in one hundred countries, and it's being adapted for U.S. audiences by Lauren Gussis, an Emmy-nominated television writer and producer and one of the geniuses behind Showtime's hit series *Dexter*.

"I think there's something so interesting about the intersection of science, which is generally a more reserved discipline, and the exuberance and passion with which Einstein approached it," Gussis told me when we spoke. "There was a fearlessness."

Her version of the show may end up including some sort of time travel. She told me she's consulting one of her distant cousins who's a theoretical physicist to make sure she's getting her science correct. But the biggest difference from the German original is a gender swap: the American version will feature the great-great-granddaughter of Albert Einstein. Gussis is working on making it a feminist narrative. "The family name gives her legitimacy and a platform to find a voice of her own, but it's really not her own. It's still because of a legacy of another man. So, she has to do the inner work to really be empowered over the course of the series."

Gussis is bringing a part of Einstein into making the show. "The fact that he does thought experiments, I was like, 'Oh my god, I do that!' That's how I write stories. That's why I'm good at running a writers' room. I always refer to it as the labyrinth. Let's just take the time and be mice and go down all the different paths. Well, if we tell the story this way, it goes like this. And if we start here, it goes like this. I've always been the person in the room who can very quickly tell you all of the divergent points based on the changes in the story. And in reading about Einstein, I realized that that's something I inherently had been doing."

My wife Elizabeth teaches a college course on the psychology of TV viewing, and she calls TV shows like this "pro-social" in that they perform a common good. Series like *Grey's Anatomy* may have a primary focus of being entertaining, but they serve a secondary purpose of teaching us about medical issues, like the treatment of patients with AIDS or the ethics surrounding a "do not resuscitate" order. This new detective show fills a similar void: Solving crimes is enjoyable, sure, but the backbone of the series allows for a fun, fictional entrée into the mind of Albert Einstein. At its heart, the series shows how he forced himself, and inspired us, to look beyond the obvious.

And that's another way that Gussis feels a kinship with Einstein, by always thinking outside the box. She has the phrase "Believe in believing the impossible is possible" tattooed on her body. "That's my entire mantra of how I operate in the world."

Around the world, Einstein is more than a mere historical figure. He is a cultural muse. A DC engineer built a robot and taught it to paint a portrait of Einstein. There's the Instagram celebrity in England who made a statue of Einstein made entirely—and cleverly—out of a candy called Smarties. People send me photos of their Einstein tattoos. There are Einstein comic books, graphic novels, and board games. Posters of Einstein sticking out his tongue line dorm room walls. There are street murals of Einstein all across the globe, including one by the famous yet elusive artist Banksy. It features Einstein carrying a sign that reads LOVE IS THE ANSWER. It's now on a T-shirt, too. "Everything about him is electric," said artist Winifred Rieber, who painted Einstein's portrait. "Even his silences are charged."

Fans of Einstein—or, for that matter, Harry Potter or Star Wars—are not satisfied to be passive bystanders to existing content; they want to be included in the creation process. And the internet provides ample opportunities for them to be a part of this participatory culture. There is a whole subculture of people who write fan fiction, including some about Albert Einstein. Others send in Einstein drawings they've created, and each week I choose the best one for a "Fan Art Friday" post on Einstein's social media pages. I once helped a magician perform an Einstein-themed trick on the popular nationally syndicated morning show *Live with Kelly and Ryan*. For so many people who can't fully grasp Einstein's equations, these pop culture touchpoints are the way they connect with him.

In a rap song, Kanye West name-drops the world's most famous genius, calling himself "this generation's closest thing to Einstein." This is not the first time, nor certainly the last, the physicist will find himself attached to a musical icon. He was one of the celebrities on the cover of the Beatles' famous *Sgt.*

Pepper album. A 1979 issue of *Look* magazine, published on the centennial of Einstein's birth, featured an odd cover duo: Elvis and Einstein. Kelly Clarkson also named a song after him, and Mariah Carey titled her $E=MC^2$ album after his most famous theory. DJ John Vader wrote a rap called "Einstein (Not Enough Time)," transforming quotes from the scientist into lyrics. (He rhymes the "speed of light" with "alright.") Brian May, the lead guitarist for the rock band Queen, holds a PhD in astrophysics and wrote a book about the Big Bang that was heavily influenced by Einstein.

Constant Speed was a ballet inspired by relativity. There's an animated TV series called *Little Einsteins* and a brand of "Baby Einstein" toys. As far as I can tell there are no Saturday morning cartoons about Newton. An entire episode of Netflix's time-traveling *Russian Doll* series attempted to explain Einstein's theory of relativity using supermarket produce. Richard Dreyfuss and Paul Rudd are just two of the many actors who have portrayed Einstein. George Clooney named his cocker spaniel Einstein.

Some of the most iconic photos of Einstein feature him looking like a Hollywood actor, wearing a rugged brown leather jacket made by Levi's—an academic from Germany sporting the most American of brands. He's even wearing it on a 1938 cover of *Time* magazine. Einstein was never one who cared about fashion, but this jacket made him look cool. It was something Steve McQueen or Robert Redford might wear. Einstein wore it all the time; its weathered look becoming an accoutrement of his celebrity. His friend, Leopold Infeld, said that "one leather jacket solved the coat problem for years." Einstein wore it so often while smoking his pipe that when Christie's auction house put the jacket up for sale in 2016, the curators pointed out that it still smelled like tobacco. The winning bid—at $146,744—came from Levi's itself, which wanted the jacket for its archives.

"Albert Einstein was a genius and an icon. This jacket is just one more example of Levi's products authentically being at the center of culture," said Tracey Panek, an historian for Levi Strauss & Co. Panek called the jacket the company's most important artifact. When it arrived at their San Francisco headquarters, she felt like she was right next to Einstein. "We opened up the crate, and the first thing you could sense was that pungent smell," she recalled. The jacket has traveled across the globe for all to see as part of various museum exhibits. But for those who can't see the original, there's another way to access this timeless piece of history. Well, at least there was. Levi's decided to produce a limited edition run of five hundred replicas of Einstein's jacket. They cost $1,200 each and sold out within days.

So why do so many people feel a kinship with Albert Einstein? Well, for starters, it likely has more to do with his celebrity, and what he stood for, than for his science. Different celebrities symbolize different things: On the one hand, they can represent the unattainable. Think about the Kardashians and their glamorous lifestyle. Their fans confer an almost godlike quality upon them. And then there are the celebrities who represent the everyman. Tom Hanks—with his aw-shucks, down-to-earth persona—epitomizes this. When Hanks became one of the first famous people to contract COVID-19 in March of 2020, many around the world felt a kinship: well, if Tom Hanks can get it, then we're all at risk.

Celebrities can be both an extraordinary person or the emblem who reflects us all. Einstein is a little bit of both. We idolize him for his intelligence, even though we don't understand things like Brownian motion or the photoelectric effect. Yet, at the same time, Einstein possesses those everyman qualities—he dresses like a schlub. He has bad hair days just like you and me. In that sense, he doesn't seem as far removed from us.

The public's relationship with celebrities is, in essence, a one-way street. Stars cannot know everyone, but everyone knows stars. Scholars call this a "para-social relationship." They are no less important than "real" relationships in that people have actual emotions invested in them. Britney Spears has forty-two million followers on Instagram where she posts mundane videos of herself dancing around the house or trying on outfits. (She also shares Einstein quotes and shows off her Einstein tattoo.) When she announced that she had lost her pregnancy in the spring of 2022, millions of her fans interacted with the post as if they were consoling a close friend.

Think about the deaths of Elvis Presley or Princess Diana or Michael Jackson. Fans around the world had real emotional attachments to these icons, as if someone they knew in real life had died. Hundreds of thousands of them turned out for those funerals. For a science journal article she was writing, my wife conducted research about the death of Robin Williams and how his fans reacted following his suicide. Their grief, mostly expressed on social media, was communal. Similarly, when singer David Bowie and actor Alan Rickman both died suddenly, at sixty-nine in the same week in January 2016, there was an out-pouring of grief and mourning online. When Supreme Court Justice Ruth Bader Ginsburg died, one woman interviewed on TV said it was like losing her grandmother. We have a strong attachment to these people whom we've never met.

Obviously, a celebrity doesn't need to die to engender such feelings. Fans scroll endlessly through famous people's social media feeds, feeling like they have a relationship with them. After all, you're getting behind-the-scenes access—seeing pictures of them inside their home, cooking dinner, or running an errand at Target. Sure, read a magazine profile about actress Kristen Bell and you'll walk away with some knowledge about her, but follow her on Instagram and you'll always be aware of what's going on in her day-to-day life with her husband and two daughters. Social scientists call this constant stream of content "ambient aware-ness." And this drip-drip-drip of information breeds an illusion of intimacy. "This is the paradox of ambient awareness," wrote science journalist Clive Thompson in the *New York Times*. "Each little update—each individual bit of social information—is insignificant on its own, even supremely mundane. But taken together, over time, the little snippets coalesce into a surpris-ingly sophisticated portrait" like "thousands of dots making a pointillist painting."

Einstein is at an interesting nexus here because he's dead but he is (or rather I am) also posting on Twitter from beyond the grave. Take it a step further: "Stardom is not a purely mercantile phenomenon imposed 'from above' by profit-hungry media conglomerates as much as it is a socially based phenomenon generated 'from below' at the level of real people who make affective investments in particular media figures," writes Gilbert Rodman, a professor of pop culture studies. "The cultural circulation of Elvis as an icon has moved beyond the power of big business to control it: today, the people who wield the most power over Elvis's public image are the millions of individuals across the globe who are his fans." We used to have communities working together to make a quilt, each one making a patch. Today, twenty million Einstein fans send in pictures they've drawn of the genius.

Graeme Turner, a professor of cultural studies, refers to celebrities as "semiotic systems embedded with cultural meanings." Einstein is a famous historical figure, a Nobel Prize winner, Time's Person of the Century. Yet his fame is unique in that it exists beyond that vacuum. Modern-day scientists are still trying to glean insights from his theories. Everyday technology that we all use—from remote controls to GPS—come courtesy of Einstein. For many others, Einstein is that funny guy on the side of our coffee mug. He exists beyond his genius. Einstein stands as a self-contained concept, a totem, hero, myth, celebrity, and rabbinic sage all rolled into one.

I was surrounded by an army of bobblehead Einsteins. Shelves upon shelves as far as the eye can see: Einstein, Einstein, Einstein.

Oh, what's that over there? Einstein. And they're all nodding their heads up and down, taunting or tantalizing me I cannot tell. I stood in a cavernous warehouse in Alpharetta, Georgia—about a thirty-minute drive north of Atlanta. This place is so big, I have to lean my head back far just to see the ceiling. I nearly felt like a bobblehead doll myself.

Go on Amazon and search for an Einstein bobblehead and you'll get back plenty of results. Most are cheap knockoffs. The only one officially licensed from the Einstein estate is made right here at the worldwide headquarters of Royal Bobbles. It took years to prototype. Artists mocked up different hairstyles. In an early version, his mane was so wild, the strands kept breaking off. Then they went too far in the other direction: Einstein looked like he was wearing a white Darth Vader helmet. Eventually, they got it right. Different outfits were attempted. The designers settled on a tan sweater, gray pants and the professor holding his trademark pipe. When they were done, they shipped it off to Jerusalem to the Albert Einstein Archives, where the caretakers of the estate held it up in the light, looked at it this way and that, and in unison bobbed their heads in approval.

Einstein is big business for the company. The doll is consistently one of its best-sellers. "I think it's popular just because everybody loves him," said Rebecca Watson, who is giving me a tour of the factory. Watson, dressed in jeans and a Royal Bobbles branded navy button-down shirt, is a customer service manager here. "He was one of the first licensees that we went after, because no matter what religion or race or politics you follow, he's loved by everyone." Sales skyrocketed so much that the company now makes different versions of the Einstein doll, including one of him playing the violin and a miniature model that can sit on top of a computer monitor. (Collect all three!) The experience inspired them to reach out to the estates of John Wayne, Alfred Hitchcock, and Elvis Presley. A svelte, young Elvis in his iconic white eagle jumpsuit, complete with rhinestones, is another bestseller. "People love all the detail," Watson told me, as we headed into a room containing a 3D printer and art supplies.

This is where they make the prototypes, but it could easily be described as any child's worst nightmare. I stumbled over a cardboard box overflowing with doll heads. On the shelves lining the walls I spotted a bunch of headless Santas, and Watson told me you can order a custom one with your face on St. Nick's body. Everywhere I looked I saw people I recognized: King Tut, Amelia Earhart, Martin Luther King Jr., Jimmy Carter, Ben Franklin, Pee Wee Herman, Lucille Ball. The Mark Twain doll basically looks like a curmudgeonly Einstein with a messier mustache. Einstein isn't the only scientist they sell. There's also Nikola Tesla, Rosalind Franklin, and Marie Curie. I hold a special place in my heart for bobbleheads: the cover of my first book featured a bobblehead Jesus.

Watson grabbed a bobblehead of Elvira, the horror movie actress known for her 1960s-style black beehive hair. "Einstein's hair was complicated, but Elvira's was so much worse," she admitted. "We had to do her bob in two parts."

I left the warehouse with goodies in hand—several Einstein bobbleheads—that now have a permanent home on a shelf in my office, which is itself lined with various pieces of

Einstein-inspired art. "You seem a little obsessed," my wife said with a hint of concern.

The bus ride from Turkey to Germany is a grueling forty-hour ride. If you happened to be taking that trip in the fall of 2016, you may have noticed a strange and rather large box lying on the seat next to a thirty-three-year-old woman. Inside was a two-foot-tall cake sculpted into the shape of Einstein—the hair, the bushy mustache, and tongue sticking out, all topped off with a frosted cardigan and a tie. In case you're wondering, the cake itself couldn't be eaten. Underneath Einstein's inedible epidermis was a combination of foam and PVC pipe holding it all together. Although, if you were hankering for a bite of the beloved genius, I guess you could have had some of his white hair, which was made out of crystalized sugar.

The bus, the woman, and the cake were all en route to Erfurt for the International Culinary Olympics, a four-day competition where bakers from forty countries were vying for the top prize. The Einstein cake was the brainchild of Inci Orfanli-Erol, an Istanbul native with no professional baking training. It took her six weeks and several attempts in her home kitchen to get the cake just right. Days before the big event, she packed it up and bought two bus tickets. It arrived safely and in one piece—and won the gold medal.

In the past, Orfanli-Erol has made cakes in the shape of the Incredible Hulk and ALF, the alien from the 1980s sitcom, but she chose Einstein for this event because he was a public figure and she wanted something that everyone would recognize. "And he's a very funny person," she said. "People were looking at the

Einstein cake and they were all smiling. I like making people smile." Orfanli-Erol turned the win into a new full-time career. She now offers classes and is hired to bake cakes for special occasions. She still works out of her home kitchen. "My husband is so happy," she said. "He likes dessert."

On Einstein's social media pages, I've shared videos of fans making everything from pancakes to pumpkins in the shape of Einstein. Believe it or not, Orfanli-Erol's masterpiece was the *second* cake in the shape of Einstein's head that made international headlines that year. Back in April, a cop-turned-cake decorator named Dawn Butler won best in show at a British bakeoff with a cake shaped like the Nobel Prize winner. "There's no more iconic figure than Einstein," said Butler, who has designed cakes for the royal family. "I love the fact that he didn't take life too seriously. I'm often referred to as the cake genius. So, it was kind of tongue-in-cheek that I would create a genius in a cake."

Many elaborate cakes are constructed with wooden or metal frames inside to help stretch the cake into a certain shape. "It was three o'clock in the morning," Butler told me. "It was sort of a light-bulb moment when I thought there's got to be an easier way than involving my husband with a drill and a saw and loads of hot glue." So, she invented the "Cake Frame," a food grade armature that can transform into any shape you want. "It was my Einstein moment."

Einstein has inspired more than just artists, musicians, bakers, and bobblehead makers. Take, for instance, the theater community, where he's found himself at the center of many productions. A 2017 play called *Friends with Albert* featured puppets. Another play, called *Einstein's Dreams*, is based on the bestselling novel by Alan Lightman. Actor and comedian Steve Martin wrote an off-Broadway play about a young Einstein at a bar with Pablo Picasso. (Elvis Presley even makes a time-traveling cameo.)

David Ellenstein has worked in theaters across the country and starred as Einstein in Steve Martin's play. It turns out, this was not his first go-round playing a historical figure. He portrayed Jesus Christ at the Crystal Cathedral, a California megachurch. "I was crucified twice nightly for a month," said Ellenstein, who is Jewish. But Einstein, he believed, was a role he was born to play. Ellenstein's father, Robert, was the first actor to portray Albert Einstein on television back in 1956. The younger Ellenstein was so inspired by the physicist that he co-wrote his own play, called *Einstein Comes Through*. It's a one-man show about an accountant who has a second job dressing up as Albert Einstein for school assemblies. Ellenstein staged a production of the show in the spring of 2021, and Jake Broder—an actor who has appeared in *The Morning Show*, *Silicon Valley*, and *How I Met Your Mother*—appeared as Einstein.

I asked them both to join me on Zoom to talk about the show. "I had such a great time researching Einstein," Ellenstein said. "I liked the idea when he talked about how a scientist doesn't need God because the scientist sees the miracle in every molecule, that everywhere you look is a delightful miracle. He is just dazzled by everything that exists."

Ellenstein also appreciated that Einstein had a good sense

of humor and loved a good joke. "One of my favorite things was one of the descriptions I read, and we included it in our play, that Einstein laughed like a contented seal." As if on cue, Broder guffawed so I could hear his rendition of the "contented seal." The laugh, he said, comes naturally. The harder part? Making sure the mustache sticks to his face.

I asked them both why they thought Einstein is still so popular, decades after his death. "He was a complicated, fascinating person whose reach was amazing," Broder said. "He's mentioned in the same breath as Gandhi and Martin Luther King Jr. He seemed to have a grasp of the interconnectedness of humanity and became a symbol of that."

Ellenstein took a more grounded approach. "Einstein is credited with being a genius and coming up with these amazing breakthrough theories, yet he seems to be so human, he seems to have such a joy," Ellenstein said. "He had a love of food and a love of women and a love of jokes. That's not how we normally think of the academic study of science. And I think that's what's so appealing and draws people to him, that he is both a lover of life and a pursuer of intellect."

Duffy Hudson, a theater actor from Los Angeles, has created a cottage industry putting on a one-man show about Einstein for school children around America—performing in California, Iowa, Illinois, Nevada, and Texas, just to name a few. He usually gets an uptick in requests around March 14, Einstein's birthday. "That month, everybody wants Einstein. And also, Houdini," he told me, referring to another character he embodies in one of seven different shows where he impersonates the likes of Edgar Allan Poe and George Burns. Occasionally, he partners up with a woman who looks like Marilyn Monroe.

Hudson was first introduced to Einstein when he was in the third grade and his teacher showed the class a documentary about the famous scientist. "Oh my god," said Hudson. "It was mind-blowing." Flash-forward a few decades and the out-of-work actor was looking for some steady income. And that's when he made an intriguing discovery: Portraying historical figures can be good business. Through his contacts in Hollywood, he booked a few gigs in advance and then got to work.

To prepare for the one-man show, which lasts about an hour, Hudson buried himself in Einsteinology—reading books, taking physics courses, and hunting through Los Angeles's vintage clothing stores in the hopes of finding the perfect cardigan. He said he barely left his house for a month. "I left to get food, and I left to go to the library to do research and to go find props and costume parts." He emerged to perform his first show on Einstein's birthday. Fifteen years and more than eight hundred shows later, Hudson summed it all up simply: "It's great fun."

He must be doing something right. He was voted the best one-man show in all of Los Angeles by the L.A. Daily News. He gives all the credit to Einstein. "I don't think he really let his ego take over," Hudson said, pontificating on the universality of Einstein. "He was genuinely all about the work. And of course, he didn't live in a time of Facebook, either. So, our opinion of him didn't get clouded. And he gave us such a gift."

Albert Einstein's story continues to be told in myriad ways each day: artists like Hudson in his performances, producers of TV shows and movies attempting to paint an accessible portrait of the world's favorite genius, and bakers and bobblehead makers looking to Einstein as their muse.

And then there's online: well, that's more of the Wild West. On Einstein's official social media accounts, I diligently check everything that I share to make sure I'm posting accurate information. Alas, I can't stop others from messing up: like Ivanka Trump did at the beginning of this book when she, like so many others, misattributed an inspiring quotation to Einstein. But as I would soon find out, Ivanka Trump was just the tip of the iceberg.

EINSTEIN IN THE AGE OF FAKE NEWS

"The search for truth and knowledge is one of the
finest attributes of man—though often it is most
loudly voiced by those who strive for it the least."

—ALBERT EINSTEIN

For all of the adoration cast upon Albert Einstein, you might
be surprised to learn that it almost didn't happen. The fame
and stature, the respect and admiration, all of it—indeed,
Einstein's entire career and his reimagining of the field of physics
and the universe as we understand it—was almost derailed by
a concerted effort to knock him off his pedestal. The tool used
by the mob hoping to take down Einstein? Fake news. And it's
something Lenny Pozner can certainly relate to.

It was election night, November 2016, and Lenny was sitting
by himself at a bar. Beer in hand, he watched the results roll in on
a nearby TV. Everyone stared mouths agape as Donald Trump
inched his way ahead of Hillary Clinton. Lenny struck up a conversation with the guy sitting next to him and asked if he had voted.
The man, who had perhaps imbibed a little too much, launched
into a diatribe: about shadowy forces that control the government
and about conspiracy theories. It didn't take long before the man

was talking about how mass shootings are really staged by the government with crisis actors, all for the purpose of getting public opinion to turn against Americans' right to bear arms.

"Staged?" Lenny asked him. "What do you mean?"

The man at the bar had an example at the ready and started talking about the 2012 shooting at Sandy Hook Elementary School in Newtown, Connecticut—the deadliest school shooting in U.S. history. The man told Lenny that he had proof it was all fake. The man had seen the same kid, a six-year-old boy named Noah, "die" in a second mass shooting in Pakistan precisely two years later, on December 16, 2014. In that incident, the Taliban slaughtered more than one hundred students. During a vigil afterward, mourners held up posters featuring children from around the world who had been victims of mass shootings. In a show of solidarity with the kids from Sandy Hook, one sign featured a photo of Noah. A perfectly reasonable explanation. But the man at the bar thought something more sinister was afoot. He knew Noah was not only an actor, but that he was still very much alive, enjoying a happy life with his family.

At this point, Lenny reached into his pocket, grabbed his wallet, and took out his driver's license to show who he was to the man at the bar. You see, Lenny is all-too-familiar with the Sandy Hook shooting. He knows for certain that it happened. Because he is Noah's father.

Lenny told me this story from an undisclosed location and made me promise not to reveal his whereabouts. For a decade, he's been pushed into hiding by an army of conspiracy theorists who have harassed him and threatened his life. So how exactly does one go from grieving father to hoax victim?

A mass shooting brings about chaos, not only to the event itself, but to the wider world in the weeks and months afterward. Those who want stricter gun laws use the opportunity to advance their cause, while others shift the blame to issues like mental illness or poor school security. In a sense, the tragedy becomes a tabula rasa for people to superimpose whatever beliefs they want onto it.

The only way Lenny knew how to cope in the aftermath of the shooting was to avoid all the noise. Besides, he was busy burying Noah and caring for his two other children who were also in Sandy Hook Elementary that day and who had narrowly survived the attack. One of them was Noah's twin sister, Arielle. Lenny saw glimpses of what was to come when a relative who did not have permission to do so established an estate trust in Noah's name. Fake charities were set up. Lenny's mail was being stolen. It was adding needless stress to his already unimaginable pain.

The Associated Press set up a camera on a hilltop near the cemetery and shared video of Noah's funeral to YouTube. "Look, this was a private moment," Lenny said when he contacted them. "You didn't have my permission to film it or broadcast it." It took a while, but he got them to remove the video. "That was probably my only good experience with the media," Lenny told me. "Everything else was a nightmare." Without an official report from the Connecticut State Police—that wouldn't arrive for more than a year—newspapers were left to fill the vacuum, often with incorrect reporting. One journalist wrote that Noah was shot eleven times—information not based on actual facts. It took Lenny months to get them to correct the record, and that was only after he met with the medical examiner and got a copy of Noah's autopsy to show the reporter. "It was just all this housekeeping stuff that needed to be done." He spent nearly a year after Noah's death trying to stamp out this misinformation. Politicians would repeat these falsehoods in speeches— that Noah was shot eleven times or that, bizarrely, Lenny was a Hasidic rabbi. In turn, Wikipedia would cite these data points because they were said by a prominent figure or reported on by the media. Lenny knew that even the smallest newspaper discrepancy about Noah's death would be immortalized for the history books.

I asked Lenny why it was so important to spend his time correcting details like the number of bullets that killed Noah. Why not just ignore it and move on? He responded—in his reserved, measured, and soft-spoken manner—with a Holocaust analogy. Most everyone will tell you that six million Jews perished during those years. But even as precise as the Germans were, it's likely it was not exactly six million people. That's a construct, a round and easy number to convey to the public the horrors that occurred. It could've been 5.7 million, Lenny said, and that small discrepancy in the coverage opens the door just wide enough for Holocaust deniers to walk in. It's a slow chipping away of the truth. For Lenny, this Holocaust metaphor is also a personal one. "My grandmother, my mother's mother, she didn't have a single relative because she was the only survivor in her family."

By the time Lenny finally emerged from the initial chaos of reporting after his son's murder, new drama had already snuck into his life. His marriage was falling apart, and his mother was terminally ill with Alzheimer's. And he was still in mourning. He set up an online memorial and posted photos of Noah. Lenny could not have imagined in his worst nightmare what happened next. Hoaxers copied the pictures for their own odious purposes and disseminated them across social media and on conspiracy laden websites. They slapped the word "FAKE" across Noah's face and, on a picture of Noah's headstone, Photoshopped the words "is not buried here" on it. One commenter asked: "Where's Noah going to die next?"

The amateur detectives went down a rabbit hole, dissecting news coverage of the event. In one instance, where CNN's Anderson Cooper was in person interviewing Noah's mom, there was the slightest blip in the video, leading many to proclaim Cooper wasn't even there and that he was standing in a studio in front of a green screen. They replayed it over and over like it was the Zapruder film. They also pointed to Noah's mom, who they believed didn't seem sad enough for it to be real. They dissected the 911 calls, claiming shots heard in the background could not possibly have come from the gun the killer was reported to have used. A YouTube video boasting to "fully expose" the Sandy Hook shooting racked up ten million views. Doubt was sowed in people's minds, and they lost confidence in institutions.

Surprisingly, Lenny understood where they were coming from. He himself had dabbled in the occasional conspiracy theory. Like maybe spaceships did land in the Nevada desert and the government was covering up. "I've always been interested in these alternative ideas," he said. "I grew up watching *In Search of*, a show narrated by Leonard Nimoy. It's mostly bullshit now if you listen to it, but I found it interesting when I was a kid." The conspiracies swirling around Noah's death may have been left to wilt in a dark corner of the web had it not been for Alex Jones.

Jones was a popular talk radio host, whose pointedly titled show *Infowars* was a place where conspiracy theories—no matter how vile—could catch fire and spread. It didn't take long after the Sandy Hook shooting for Jones and his audience, self-described gun-toting patriots, to invent a new reality about the tragedy. They said Lenny was being paid by the government as part of an elaborate plot to confiscate their firearms. Jones and his followers re-shared the doctored images of Noah to help fuel their false narrative. Lenny offered up Noah's death certificate and autopsy report. But that only fed the conspiracies more, as they claimed the documents were counterfeit. They told Lenny the only way they would believe him is if he exhumed Noah's body.

Lenny was an IT consultant with more than a working knowledge of how the internet operated. More than that, he was an aggrieved parent, suffering the unfathomable dual tragedies of losing a child and countering a public narrative that his son didn't actually die. "I'm going to have to protect Noah's honor for the rest of my life," he said. He set about correcting the record yet again. He spent countless hours each day flagging posts that were using photos of Noah—on YouTube, on Twitter, on Facebook, on blogs and websites. Instead of engaging in the

muck of the conspiracy itself and trying to explain to these platforms that his son was in fact murdered, Lenny took a simpler
approach. When he would report an item, Lenny simply said
that the poster was using Noah's photo without his permission
and that he wanted it removed. It was not an issue over a hoax
or fake news. His legal strategy was novel: the information may
be wrong or mendacious; but it was also, at its core, copyright
infringement. Family photos of Noah with his sisters at the
beach, playing in the park, or sliding down the stairs—Lenny
owned that material. He had found a technical loophole to this
specific disinformation campaign. Ever so slowly, thousands of
posts began to get taken down. This only enraged the conspiracy
theorists who called their posts' removal "Getting Poznered."

Alex Jones—whose audience doubled between 2013
and 2016—told his listeners that Lenny was taking away people's free speech by having their posts removed and, in some
instances, even getting entire websites taken down. Jones called
it a "war" on censorship and mobilized his millions of fans to act
accordingly. Almost immediately, they started the harassment,
finding Lenny's personal details and releasing them online—
his social security number, his phone number, and his home
address. He was able to get some stuff taken down due to pri-
vacy violations—like a YouTube video in which Jones shared a
Google Earth satellite photo of Lenny's home. Lenny moved,
and then his new address was revealed. So, he kept moving—
more than a dozen times. They left voice mail messages, some
with an added dash of antisemitism, threatening to kill him.

Lenny was just the latest—if perhaps one of the most

appalling—victims of fake news. But he was by no means the first. As we'll soon discuss, Albert Einstein's career trajectory was nearly derailed due to an orchestrated disinformation campaign against him.

It's easy to think fake news is a modern invention. Hackers in 2013 took control of the Associated Press's Twitter account and posted the headline: "Breaking: two explosions in the White House and Barack Obama is injured." The Dow plunged after the news broke and a stock market crash resulted in a loss of $136.5 billion. A few years later, a man walked into a Washington, DC, pizzeria armed with a semiautomatic rifle. He said he was there to rescue the child sex slaves in the basement, something he had been made aware of because of fake news stories circulating online in far-right message boards. A 2019 survey by the Pew Research Center found that 50 percent of Americans think made-up news is a significant problem for the country—more than other issues like climate change (46 percent) and racism (40 percent). But it would be wrong to think that fake news is a problem created by our social media–driven society.

Fake news did not begin with the advent of the internet. Indeed, the founding of America happened, in part, due to fake news. During the Revolutionary War, when the country was divided over its attitudes toward the British, a polarized press corps filled the gap with partisan news. Some newspapers attempted to undermine British authority by publishing highly exaggerated stories of corruption by British officials.

The articles, which incited hatred toward the British, were often penned anonymously—some historians believe future U.S. President John Adams had authored a few of them. In 1770, during a confrontation in Boston, a mob of hundreds swarmed around a small group of British soldiers. In an attempt to save themselves, the British soldiers killed five people in the crowd. Artist Henry Pelham painted a picture to immortalize the event. But Paul Revere, the American revolutionary, saw an opportunity to produce patriot propaganda. He copied Pelham's painting, but importantly changed key elements, including depicting a British captain giving orders to fire at unarmed citizens. Revere called his version "The Bloody Massacre" and offered prints of it for sale a week before Pelham could get his to market. My wife, whose academic research includes the advent of fake news, hung up Revere's version of the painting in our living room.

Benjamin Franklin may have been the most egregious peddler of fake news in the early days of our republic when, in March 1782, he printed a hoax supplement to the *Boston Independent Chronicle*, complete with fake news and fake advertisements. One article detailed a horrific crime perpetrated by the British in which they coordinated with Native Americans to scalp seven hundred "defenseless farmers, and women, and children." Except, of course, it wasn't true. "The form may perhaps not be genuine," Franklin told his friends, "but the substance is truth."

By the end of the nineteenth century, "yellow journalism" was on the rise—a style of reporting that had, well, very little basis in actual reporting. Circulating disinformation is nothing new; but its targets certainly were.

Albert Einstein would likely not at all be surprised about our current state of fake news. He himself was the target of politically motivated disinformation. I decided to call up Matthew Stanley, a professor of the history of science at New York University, who wrote an entire book about how Einstein defeated bigotry and nationalism while trying to prove his general theory of relativity.

While World War I was raging across Europe from 1914 to 1918, Einstein was practically alone—he was separated from his first wife and many of his scientific colleagues had been enlisted into the war effort. He wrote letters to his friends about how frustrating it was that otherwise intelligent people were accepting what seemed to him to be obvious untruths about the war. A group of ninety-three German scientists, including six Nobel Prize winners, penned what came to be known as "The Manifesto of 93," tying their scholarship to the jingoistic militarism of their country. "His thoughtful friends had suddenly become patriotic sheep," Stanley wrote in *Einstein's War*. "Rationalism had been displaced by nationalism." Scientists from Britain, now deemed the enemy, were summarily removed from German scientific journals.

These days, it's okay if we disagree with the political beliefs of a crazy uncle, because we really only have to deal with him once a year at the Thanksgiving table. And yet Einstein, an avowed pacifist, realized he still needed his colleagues' intellectual camaraderie, the personal connections to science. "My sense of it is, that he had so few close friends, that he'd do anything to keep them even if it meant having to deal with this kind of stuff," Stanley told me.

But Einstein saw the seeds of what he called "collective insanity" planted in his friends' minds. "Einstein talked about what nowadays we would call groupthink," Stanley explained. "He called it the herd mind, that people just go along with whatever is being told to them and don't think deeply about it. He blamed a lot of the world's ills on that kind of top-down propaganda."

Einstein hoped this would cease by the end of World War I, but that notion would be quickly dismissed. After the war, an anti-relativity society formed, led by fellow scientists and former friends—Philipp Lenard and Johannes Stark, both German physicists. The group printed pamphlets that discredited Einstein and his novel theory that, as Einstein described it, drew "as much respect as it does suspicion." It was more than the fact that Einstein was upending physics and the status quo. Yes, there was certainly that. But much of the criticism was laced with antisemitism. The pamphlets were circulated in an ad hoc way—one person passing it along to a friend who, in turn, would forward it to someone else. In the present era, you might have found them all congregating in the same Facebook group.

Einstein was puzzled by the whole endeavor and showed up to one of their events. Nobody seemed to notice he was there: At the door, he was handed an anti-Einstein pamphlet. "He thought that these people were actually not very dangerous because they're so silly and so poorly informed," Stanley said. "He thought it's all faintly ridiculous."

But Einstein would eventually change his tune. When his

general theory of relativity was finally proven during a solar eclipse in 1919 (which we discussed in the earlier chapter about alien research), Einstein suddenly found himself the most famous Jew in the world. "After the war," Stanley explained, "the right wing is looking for convenient targets, and Einstein is just too good a target for them to pass up. The fact that his science is so weird is a helpful hook to these things."

As the Nazis rose to power, the German government supported this disinformation about Einstein. By this point, there were two separate strains of falsehoods: One was that relativity was outright wrong. On a trip to the Einstein Archives, I found a copy of a 1931 pamphlet called *One Hundred Authors against Einstein* filled with anti-relativity essays. Simultaneously, there were groups saying that relativity was the greatest thing in the world, but that Einstein had stolen the idea for it from German and Austrian scientists. Einstein was targeted as enemy No. 1, a bounty was placed on his head, and he ultimately fled Germany. "Blind obedience to authority," Einstein said, "is the greatest enemy of the truth."

Lenny Pozner wanted to take back control of the narrative. He was tired of being hunted and harassed. But he was just one man determined to get the truth out from under a mountain of lies. Twitter has more than eight hundred million new tweets sent every day. Facebook has nearly three billion active users per month. YouTube has five hundred hours of video uploaded to its servers every minute. Each platform has architected systems that are nearly impossible to moderate. "I knew what was right. I knew what was wrong," Lenny told me. "I knew what I needed to do."

So, Lenny launched the HONR network, a non-profit aimed at providing assistance for victims of mass casualties who are targeted by online hate. He enlisted the help of dozens of volunteers—including some people who had originally believed Sandy Hook was a hoax but who were now clear-eyed and repenting for their past sins.

Since then, Lenny has seen some success with Facebook taking down Sandy Hook misinformation. "Removing an entire group on Facebook, the volume of material that disappears is phenomenal," he said. "Removing a user account that's existed for a decade, everything disappears. Everything that person commented on, everything that has been re-shared that that person posted, it all disappears." Now, Lenny said, he's worried about Facebook's future plans to create a metaverse, a hybrid of the real and online world. There, he said, "the potential for human injury is probably far greater."

Working with those tech platforms was just the start. Next up for Lenny was targeting individuals who were harassing him. Some of the trolls, like the one who harassed him with voice messages, were arrested. When a man in Australia made repeated death threats, Lenny filed a report with that country's federal police. A woman in Florida was sentenced to five months in prison for threatening him with death. Lenny got a professor who blogged falsehoods about Noah fired from his tenured position. Lenny sued the editors of a book called *Nobody Died at Sandy Hook: It Was a FEMA Drill to Promote Gun Control.* A judge awarded him $1.1 million in the defamation case and the rights to the book.

And then Lenny went after the match that was lighting all these fires—Alex Jones. He was among the first people to sue Jones in his home state of Texas. A month later, other Sandy Hook families sued Jones in Connecticut. Under oath during a deposition, Jones testified that Noah and nineteen other kids did, in fact, die that winter morning at Sandy Hook. Lenny said he considered that public admission a victory. "I've already won."

There were four cases brought against Jones by the Sandy Hook families, and Jones lost all of them. The judge ordered him to pay damages to the families. And there is potentially a lot for them to receive: According to the *New York Times*, *Infowars* brought in more than $50 million annually from 2016 to 2020. Although, by that point in time, every major distribution platform—from Facebook to YouTube to Apple Podcasts and more—banned Jones from their services. In July 2022, the parent company for *Infowars* filed for bankruptcy.

Lenny and his HONR network have chalked up more wins—they've got some tech platforms to make victims of tragedies a protected class. In much the same way it's considered hate speech to bully and harass Jews or Black people or members of the LGBTQ community, it's now against the rules to bully the families impacted by mass shootings.

As Lenny knows all too well, online harassment can have real-life consequences. A Georgia poll worker, Wandrea Moss, testified before Congress that she was forced to go into hiding after Donald Trump and his lawyer Rudy Giuliani baselessly accused her of rigging the 2020 election, inspiring a conspiracy-fueled online mob to go after her. "I don't go to the grocery store at all," she said. "I haven't been anywhere at all. I've gained about sixty pounds. I just don't do nothing anymore." A conspiracy theorist showed up at the home of Moss's grandmother and said they were there to make a citizen's arrest. "It's affected my life in a major way—in every way," Moss said. "All because of lies."

Albert Einstein moved to the United States in 1933 where he was, by all accounts, welcomed by adoring throngs. But fake news followed him from his native Germany, although this time with less nefarious purposes. "He was this sage," Professor Stanley explained, "and everyone wanted to know his opinion on everything." People would ask Einstein his opinion on modern art, on jazz, on God. When a small-town reporter saw Einstein on a train and asked him a question, he would simply say whatever he thought. "This enormous pile of weird sayings is attributed to him, some of which are true, and some of which are not," Stanley said. So, it became easy for people to misattribute quotes to Einstein—like Ivanka Trump did at the beginning of this book.

Einstein's name got attached to all sorts of random state-ments and ideas, a distinctive feature of today's internet. Stanley believes that Einstein would completely grasp the current iter-ation of fake news. "He saw this happening back in 1915. His German friends believed all the stories about British atroc-ities, but they never believed any of the stories about German atrocities." Einstein understood that was tribalism run amok. In the modern-day parlance it's an echo chamber, what internet activist and author Eli Pariser dubbed "filter bubbles." We choose which news sites we read online, often opting for ones that agree with our worldview. Moreover, on social media, if we are only friends with people who agree with us, it creates a distorted sense of reality. This problem is made even worse by algorithms that deliver personalized news feeds based on what you've interacted with before. When you watch one conspiracy theory video, YouTube quickly recommends another. The more you click on a particular news source—say, Fox News—the more Google will serve you results from Fox News. So, your search for "Trump election lies" may look a lot different from that of your neighbor.

"Einstein gets this up-close experience with conspiracy theories in the 1920s and '30s when people were looking for scapegoats and ways to explain terrible events in terms that are comfortable to them, with social and political explanations that reinforce their own worldview. And then the herd mind takes over and everybody starts agreeing with their friends," Stanley said. "So, I think he would find it tragic, of course, but he would recognize it instantly."

Einstein certainly spoke out about things to which he had an impassioned opinion—whether it was about the rise of Hitler or the scourge of racism in the United States. But he didn't go through, point by point, refuting the disinformation that he saw. Lenny has a similar approach with people who are too deep into conspiracy theories. "You can just make the information available to them, so that if they're ever ready they'll find it," Lenny said. "You can't convince a Buddhist to become a Muslim. You just can't. But as long as the Quran is available somewhere for them, when they become interested in it, they can find it."

Einstein and Lenny's strategy would appear to be a smart one, as research shows that trying to disavow people of their beliefs is an uphill battle. "What does change people's minds is providing a different narrative, a whole different alternative structure that people can latch on to," Stanley said. "That is something that Einstein was pretty good at doing. There's some sense of leading by example." During the civil rights movement, for example, Einstein didn't refute segregationist ideology point by point. Instead, he simply went to an historically Black college and taught some classes. "It was just showing that people of color could think as deep thoughts as anybody else. They could understand relativity perfectly." Throughout his life, Einstein always sought the simplest, most elegant, explanation to complex problems.

I told Stanley about my work with the Einstein social media accounts. "I don't know if Einstein would have been on Twitter," he said. "But I could imagine him having a blog that everybody else would retweet." Reporters would follow Einstein around all day, asking him odd little questions. And he was really good at coining a phrase. So, people would write down whatever he said, and send it around. In that sense, any of Einstein's quotes could be tweets. "One of the reasons his quotes are so compelling is that they have this oracular quality where you have to read something into them."

Einstein was particularly adept at conversing with the press. In addition to the wise quotes, Stanley said, "he's photogenic in that very distinct particular way, and is willing to play into that. He sticks out his tongue, he throws his hat up in the air, so people can take pictures of it." This kind of celebrity behavior may feel pretty standard today, but back then it was a bit radical. "He was good at it. Charlie Chaplin was good at it. But Marie Curie, for instance, was not. She refused to do that kind of thing. So, nobody quotes Marie Curie."

Einstein's fame coincided with the rise of mass media in the twentieth century—newspapers, radio, television. Each medium had its own problems with fake news. In print, as we brought up earlier, Benjamin Franklin played a role. On radio, there was the infamous *War of the Worlds* broadcast, which confused listeners into thinking an actual alien invasion was taking place. Neil Postman, in his groundbreaking 1985 book *Amusing Ourselves to Death*, wrote that TV news "creates the illusion of knowing something but which in fact leads one away from knowing." News reports from stentorian figures like Edward R. Murrow and Walter Cronkite would eventually devolve into the morass we see twenty-four hours a day on cable news.

The internet, with its algorithms that value engagement over truth, has taken all of this to an entirely new level. Disseminating misinformation has become so much easier. It's no longer just a crazy guy on the subway shouting conspiracy theories. Now, everyone—even Joe Schmo in his parents' basement—has a megaphone that can reach millions of people within minutes. "In the pre-internet era, disinformation was as difficult and expensive to produce as truthful information," wrote Richard

Stengel in the *New York Times*. "You still had to pay someone to do it—you still had to buy ink and paper and distribute it. Now, the distribution cost of bad information is essentially free, with none of the liability of traditional media."

When a person tells their followers on YouTube about a particular news event—like the Sandy Hook shooting—they are, in essence, separating the content from its source. This makes it easier for information to be misinterpreted or spun for ulterior motives. When speaking of the invention of the printing press, science historian George Dyson describes this concept as when "knowledge began freely replicating and quickly assumed a life of its own." There is a wealth of readily available information at our fingertips, just a quick Google search away, which means we can consume more information about more things, without ever having to speak to someone who was actually at a particular event.

And you don't have to be a conspiracy theorist wearing a tin foil hat to believe a piece of fake news. A study out of Princeton University found that Facebook played a central role in disseminating content from untrustworthy news sites in the months leading up to the 2016 presidential election. More than a quarter of adults visited a fake news site during that time. An analysis from Buzzfeed from that same period found that the top twenty fake news stories had a higher engagement on Facebook than the top twenty real news stories.

Lenny said the Sandy Hook misinformation was "the canary in the coal mine, because it was the intersection of all of these things that were going on in society with technology and media." Conspiracy theorists now had so many new tools at their disposal. They could sift through the Facebook profiles of teachers at Sandy Hook, or of any resident in Newtown for that matter, scroll through their list of friends, zoom in on photos, see seemingly intriguing things, and attempt to draw conclusions from mere coincidences. None of this existed in 1999 when a mass shooter killed twelve students and one teacher at a high school in Columbine, Colorado. "That decade between 2010 and 2020 was when social media was really stress tested, because nothing existed before that," Lenny said. Sandy Hook served as a foundational story that predicted so much of what followed.

We have tumbled, some would argue, into a post-truth world. As Kellyanne Conway, an adviser to President Trump famously said: there are alternative facts. After Sandy Hook, it seemed, there was an uptick in the consistency of these types of truth denialism: after other mass shootings, with the coronavirus pandemic, and leading up to the rewriting of history at the outcome of the 2020 election, with claims that Joe Biden didn't win. Those lies of a stolen election ultimately led to an attempt to overthrow the American government on January 6, 2021. "The first casualty of war is truth," said Elizabeth Williamson, a reporter for the *New York Times* who wrote a book about Sandy Hook.

Lenny mentioned the coronavirus vaccine and how it became politicized, with some vaccine skeptics taking advice from people online who never went to medical school. "If there's one article that says someone got a blood clot in their lungs, and they took the vaccine, that's all the proof they need," Lenny said. "That somehow goes against the tens of millions of people that may have been saved. There's nothing you can do to talk people out of that. They just need to learn the hard way. And I think that those people existed, they always existed."

In Einstein's later years, the conspiracy theories still hounded him. He was often critical of the U.S. government—whether over his new country's warmongering or in the way it treated its Black citizens. Some thought Einstein was a secret Russian spy, an accusation that took on added weight during the era of McCarthyism, when there were increasingly hostile attitudes toward scientists—especially those who were foreign-born. Senator McCarthy's staff created bogus documents and doctored photographs to advance their means.

William Frauenglass, a Brooklyn high school teacher, was called before McCarthy's Senate subcommittee for a class he taught about easing interracial relations, something the committee deemed "against the interests of the United States." Frauenglass asked Einstein for a letter of support, to which Einstein, a fierce defender of civil rights, happily agreed, even allowing the *New York Times* to reprint the letter—which it did on June 12, 1953, on its front page. "The reactionary politicians have managed to instill suspicion of all intellectual efforts to the public by dangling before their eyes a danger from without," Einstein wrote. Einstein's advice to Frauenglass? "Refuse to testify."

McCarthy wasn't thrilled with Einstein's letter and dubbed the Nobel Prize winner an "enemy of America" and "a disloyal American." Einstein wasn't deterred. In a letter the following year—in March 1954—he pushed for "the right to search for truth and to publish and teach what one holds to be true." Like Lenny, Einstein was harassed by people who believed the lies. One woman wrote to Einstein's employer saying that she had "no patience" for him. Another critic was more direct: "I suggest he move to Russia—and soon! We don't need him."

The FBI became interested in Einstein as far back as 1932, when a concerned group calling themselves the Woman Patriot Corporation authored a sixteen-page letter to the State Department trying to convince them to not let Einstein immigrate to the United States. (Einstein became a U.S. citizen in 1940.) The FBI's monitoring of Einstein picked up steam around February 12, 1950, when the scientist went on national television, with former First Lady Eleanor Roosevelt, to speak out against America's plan to build a hydrogen bomb. "The annihilation of any life on Earth has been brought within the range of technical possibilities," he warned. The very next morning, FBI Director J. Edgar Hoover asked his agents for all "derogatory information" they could find about the scientist, according to Fred Jerome, who wrote a book about the FBI's Einstein file.

The surveillance—going through Einstein's trash and mail and listening in on phone calls—continued for the next five years, and coincided with the peak of anti-Communist, Red-scare hysteria. Hoover kept the operation secret and sought to ensure there were no leaks. "If it got out prematurely that he was investigating the world's most admired scientist and America's most famous refugee from Nazi Germany, he knew that he and his FBI—and quite possibly the entire United States government—would face a storm of international outrage and derision," Jerome wrote. (Jerome was familiar with the topic; his family was also accused by Hoover as being Communist spies.)

By the time Einstein died, in 1955, the FBI had a file on Einstein that bulged to 1,427 pages. Peruse through those documents, which the FBI finally made available online after enough people requested access to them, and you'll find all sorts of conspiracies and insinuations about Einstein. He was a radical and a nonconformist with allegiances to other countries. He was part of a Communist conspiracy to take over Hollywood. One page said he was part of a project that used flying saucers to help Russia spy on the United States, while another said he was hard at work creating a death ray "which could be operated from planes to destroy cities."

It was a concerted campaign to undermine the reputation of Einstein—one he likely shrugged off. "He was largely indifferent to the risk of assassination by the Nazis in 1933," said Andrew Robinson, who wrote a book about Einstein and politics. "I don't think FBI surveillance would have frightened him."

After speaking with Lenny for more than an hour, he had an insight that stopped me in my tracks. We were talking about Sandy Hook and fake news and Einstein, and he looked at me and said, "Truth has now become a relative concept." He pointed out that the theory of relativity teaches that there is no truth about time—it all depends on where you're standing. Somebody on Earth experiences time differently than somebody in space. "I'm not that great at physics," Lenny admitted, "but Einstein's experience was that if you're observing time from Planet A, and someone else is observing time from Planet B, and they're looking at an event going on in Galaxy C, the observers are going to see those things happening differently.

"Truth," Lenny said, "depends on where you're standing. If you are a Republican, then truth to you is seen in conservative media, about how vile the Democrats are." Lenny and I are both Jewish and he brought up the concept of the Torah. "Who's to say our truth is more significant than other people's truths? We're seeing different things depending on what our vantage point is."

Despite Lenny's Herculean efforts, disinformation campaigns are not going away. "Disinformation," wrote Richard Stengel, "flourishes in times of uncertainty and divisiveness." In the lead-up to Russia's war with Ukraine in the spring of 2022, President Vladimir Putin told his people that Ukraine was full of Nazis, a baseless claim. Nonetheless, Putin used that as an impetus to invade, but Russians began finding out the truth from other sources—ironically from social media, which can so often be the source that spreads such lies. Putin, hoping to inoculate his citizens from outside influence, banned Facebook and other platforms in Russia.

Misinformation has been around for centuries, with countries trying to push their own self-serving narratives. But what's different now is the sheer speed and scale of how it can spread.

"Propaganda has become the next nuclear threat," said Scott Galloway, an NYU professor and host of a popular podcast about business and technology. Social media sites favor videos that engage and enrage. If TikTok keeps serving up videos of people saying the 2021 coup attempt and riot on the U.S. Capitol was merely a minor protest, it can start to change people's perceptions about historical facts. "Small tweaks to the algorithm," Galloway pointed out, "and you have what is potentially one of the most dangerous and damaging weapons in the history of mankind."

It used to be only a foreign adversary like Russia or North Korea could sow nationwide discord. Now anybody with a smartphone can do it. But experts see some glimmer of hope. Statistics show that the newest category of social media users, Generation Z and younger, who are native to the internet, are more skeptical of misinformation.

What would Einstein, who was driven by a lifelong curious sense of discovering truths about our universe, think of where we ended up? Of all the user-generated and unmoderated content? Of the deluge of false news stories and incendiary tweets? Of the elevation of everyone as an expert pundit, a genius in their own mind? To our own detriment, the internet has flattened information—truth and lies—so that it all looks the same. A 2022 survey by the polling firm Ipsos found that 34 percent of Americans say they're skeptical of science and 47 percent believe science divides people with opposing views. "Addressing widespread misinformation is important," said Dr. Jayshree

Seth, who was involved in the study, "because 81 percent of Americans say that if we cannot trust news stories about science, there will be negative consequences to society," like an increase of the impact of climate change and more public health crises.

Einstein, a champion for truth and knowledge, would likely be a proponent of modern-day media literacy campaigns, efforts to train people how to be an informed news consumer amid a veritable tidal wave of content. Take a look at the article you're reading online: Is it from a trusted news source? If it is, check the date—is this a story from today or three years ago? My wife, who teaches college students on this topic, offers up a few possible solutions for news consumers: encouraging them to spend more time engaging in "lateral reading" by which they read multiple sources for the same news story; technological fixes like algorithms that can detect fake news automatically; and social media designs and interventions that might hinder the sharing of fake news. In 2020, in an attempt to slow down the spread of fake news, Twitter tested a feature that asked people if they had read more than just the headline of the article they were about to share. Regardless of these efforts, at the end of the day there is no cure for fake news.

"Information is flowing at the speed of light," Lenny said to me, taking a cue from Einstein's theories. "So, everything is amplified, everything is accelerated. At some point, we'll reach a critical mass where there's so much disinformation that there will no longer be any truth. It'll just be different opinions." As Einstein himself once said, "The problem begins with whether truth is independent of our consciousness."

Part of the blame, of course, lies with us. After all, we as a collective species are the ones sharing information on social media without first even reading the article. We're the ones choosing to spend hours each day, devoting our time and attention, scrolling through our endless feeds. But we can try to do something: In my own small corner of the web, I take personal responsibility for what I post to Einstein's twenty million followers on social media. The last thing I want to do is post fake news to Einstein's Twitter page. I make sure I'm posting science articles from reputable sources because I am guided by a principle from Einstein, who said, "Whoever is careless with the truth in small matters cannot be trusted in important affairs."

Perhaps the tide is already turning: A 2022 study of more than ten thousand people from twenty-four countries found that when shown statements from either a spiritual guru or a scientist, the majority of the participants—no matter their level of religiosity—gave credibility to the scientist over the spiritual guru. The researchers wrote that their findings suggest that "across cultures science is a powerful and universal heuristic that signals the reliability of information." They dubbed their discovery the Einstein effect.

PRESERVING EINSTEIN

"The Hebrew University in Palestine will
become a new 'holy place' to our people."

—ALBERT EINSTEIN

After Albert Einstein died, Israeli officials came to clean out his office at the Institute for Advanced Study in Princeton. They loaded up all of his papers, correspondence, photos, medals, and other ephemera—between there and his home at 112 Mercer Street, they counted tens of thousands of items. They left the furniture for the Historical Society of Princeton. The items were boxed up into big wooden crates and loaded onto a truck, and then a plane, and then a van, where it traveled 5,736 miles to the campus of Hebrew University in Jerusalem. There were police escorts along the way.

The reason for this pomp and circumstance was because Albert Einstein, an ardent Zionist, was one of the founders of Hebrew University. In fact, Einstein's first trip to the United States, in the spring of 1921, was part of a two-month tour to help raise money for the establishment of the school. Einstein was invited on the trip by Chaim Weizmann, a fellow scientist, who would eventually become the first president of the State of Israel.

Already famous for proving his theory of relativity and on the cusp of winning that year's Nobel Prize, Einstein was greeted throughout America with "the mass frenzy and press adulation that thrill a touring rock star," wrote Walter Isaacson in his biography of Einstein. "The world had never before seen, and perhaps never will again, such a scientific celebrity superstar." While in the United States, he lectured at Columbia University and the National Academy of Sciences. Princeton gave him an honorary degree. He went to the White House and met with President Harding. When enough money was raised for Hebrew University, Einstein joined its Board of Governors and gave the first scientific lecture. After the school opened, he remained involved for the rest of his life despite living in New Jersey. As someone described it to me, Einstein was "a provost acting from exile."

"Einstein lent the *prestige mondial* of his great name, and in fact gave his heart, to the movement which created the state of Israel," wrote Isaiah Berlin, a British philosopher. "No Zionist with the least degree of self-esteem can refuse to pay him homage if the opportunity of doing so is offered to him." Years later, as Hitler rose to power in Germany, Einstein wrote that "Palestine is not primarily a place of refuge for the Jews of eastern Europe, but the embodiment of the re-awakening of the corporate spirit of the entire Jewish nation." He felt a moral obligation to support the country.

Weizmann died in the fall of 1952, and Israeli Prime Minister David Ben-Gurion thought he had the perfect person who could step into the job. "There is only one man whom we should ask to become the president of the state of Israel," Ben-Gurion told his political secretary, Yitzhak Navon. "He is the greatest Jew on Earth. Maybe the greatest human being on Earth. Einstein. What do you think?"

Telling the story decades later, at a 1979 symposium in Jerusalem marking the centennial of Einstein's birth, Navon recalled he was shocked by the suggestion. "President of the state of Israel?" I asked. "Why not head of scientific research?" But Ben-Gurion insisted.

Abba Eban, an Israeli diplomat, officially offered Einstein the job, saying it "embodies the deepest respect which the Jewish people can repose in any of its sons." But Einstein didn't want to move out of Princeton and, besides, he didn't see himself as a politician. "All my life I have dealt with objective matters," wrote Einstein in his letter rejecting the job. "Hence, I lack both the natural aptitude and the experience to deal properly with people." However, he said he was "deeply moved" by the offer, especially considering "my relationship to the Jewish people became my strongest human tie." In a letter to a friend, Einstein expressed regret and said he "was deeply touched by the offer from my Israeli brothers," adding that "it is quite true that many a rebel has in the end become a figure of respectability, even a big shot; but I cannot bring myself to do so."

Isaacson recounts what happened next: "Two days later, when Ambassador Eban ran into Einstein at a black-tie reception in New York, he was happy that the issue was behind them. Einstein was not wearing socks."

Nevertheless, Einstein wanted to leave behind some sort of legacy that would help his Israeli compatriots. When crafting his last will and testament in 1950, he decided to bequeath his "manuscripts, copyrights, publication rights, royalties...and all other literary property and rights, of any and every kind or nature whatsoever" to Hebrew University, the school he helped establish. Einstein was way ahead of his time: Before the Beatles and Elvis, before Michael Jordan got a shoe deal with Nike and Dolly Parton built an amusement park, Einstein understood the power of celebrity. Now, anytime someone buys an Albert Einstein bobblehead, Hebrew University makes a few shekels. Over the decades, the royalties have raked in millions of dollars for the school. Presumably the amount could have been even more, but the university is careful to protect Einstein's image, refusing to endorse products like Einstein vodka.

As Albert Einstein lay on his deathbed in a room at Princeton Hospital in 1955, he was in the middle of writing a speech that he was planning to give the following week for Israel's Independence Day. The address was supposed to be televised on ABC, NBC, and CBS and simulcast across radio stations. Ambassador Eban told him that sixty million people would hear it. "So," Einstein retorted, "I shall now have a chance to become world famous."

But, alas, Einstein died before he could deliver it. Part of a draft of the speech was found by his bedside.

The pages containing that entire speech now reside in a climate-controlled room at the Albert Einstein Archives at Hebrew University in Israel, the largest collection of Einstein related material in the world. There, just a handful of people—including a physicist, a grandma, an artist, and a Hasid—are working to preserve the past by curating a coherent narrative of Einstein's life. And now that staff includes me, tasked with sharing the story of Einstein to a whole new generation online.

In my everyday life, I can be quite stupid. The first day I got my new riding lawn mower, I promptly proceeded to drive it over a huge tree trunk and broke it. There are bulbs in light fixtures in my house that I still can't quite figure out how to change. I once bought a bookcase at Target and impressively put it together all by myself, only for my wife to inform me that I had built the whole thing upside down. I've been eating solid food for decades, yet I still have no clue how to boil an egg.

The irony is that in my day job I need to pretend to be the greatest genius mankind has ever seen. When I'm sitting with my laptop on the front porch, staring at our flock of chickens aimlessly zigzagging around the yard like zombie fowls, I log into Albert Einstein's Twitter account and post an article about how Stanford University physicists are trying to build space-time out of quantum particles. On Facebook, I reply to someone's comment about dark matter. On Instagram, well, let's be honest, on that platform I can share a short video of Einstein tossing his hat into the air, and watch the positive reactions roll in.

People often ask how I got this job, of becoming Albert to his tens of millions of fans. They ask if I'm a scientist or had a special connection to Einstein. What I usually say is that, like them, I grew up admiring Einstein the man, this symbol of smarts. And then I tell them it all began with an auction.

At the time, I was reporting for a news website where I had the flexibility to write on whatever topic piqued my interest: One day, it was a story about America's most irrational man. Another time, it was interviewing a guy who launched a dating app for dogs. Then one morning, I heard about a dozen letters that Einstein wrote to his friend that were going up for auction. It was a slow news day, so I published an article. My editor had this smart mantra: spend the morning writing your story and spend the afternoon promoting it. After all, you could write a Pulitzer Prize–worthy story, but what good is it if nobody reads it? (Granted, my article about a Giant Schnauzer who met his girlfriend at the dog park wasn't turning me into Woodward and Bernstein.) My editor asked me to look on Facebook to see if there were any Einstein fan groups where I could post the article. I typed "Albert Einstein" into the search bar and, lo and behold, up came the "official page of the world's favorite genius." That's literally what it says at the top. Fortunately, Facebook makes it easy for you to message the owner of a page, so I clicked the button and wrote the following:

"Hey! I'm a big fan of Albert Einstein and just wrote this article about an auction of some of his letters. I thought you might be interested in it. Feel free to share it with your followers!"

Less than five minutes later, I saw those three dots indicating that Albert Einstein was typing something back to me…

"Wow! My name is Tony and I help run Einstein's social media accounts. We're always looking for good content to share. Thanks!"

Tony posted the story to Einstein's twenty million followers. The instant deluge of traffic to my article caused our news site to crash. Needless to say, my editor was thrilled and suggested that I write more articles about the physicist. My next award-winning story about cats who wear baby clothes would have to wait. I set up a Google Alert for "Albert Einstein" and, each day, news items about him would land in my inbox. I thought maybe I'd find one or two stories a month to write about. But to my surprise, Einstein was proving more prominent today than ever. There were new discoveries based on Einstein's research or the new TV show about Einstein's life. There were so many artists making paintings of Einstein, I could interview one every week.

I kept writing all of these articles and sending them to Tony. Eventually, Tony and I started working together to come up with story ideas before I even started writing. We had weekly phone calls about Einstein.

"Einstein's birthday is coming up," Tony told me one day.

I composed a list of ten celebrities who shared a birthday with Einstein. Who knew Billy Crystal and NBA legend Stephen Curry were so astrologically aligned with the great scientist? Over the years, Tony and I became friends and I traveled to New York to visit him, and he came here to West Virginia. When Tony got engaged, he invited me to the wedding.

Tony had the keys to one of the world's most influential social media accounts because he worked for a company in New York called BEN Group, Inc.—short for Branded Entertainment Network. The firm manages the personality rights of a telephone book's worth of dead celebrities in addition to Einstein: Muhammad Ali, Charlie Chaplin, Thomas Edison, Andy Warhol, and the Wright brothers. From his social media dashboard, Tony could very easily accidentally post an Einstein picture to Steve McQueen's Facebook page. The archivists and curators who work at the Albert Einstein Archives are mostly academics; they help find material for other academics doing research about Einstein. Tweeting is not really their expertise. So, they hired Branded Entertainment Network to take care of Einstein's social media presence.

But after a couple years, Tony, a law school graduate, was looking to spread his wings and find work in his chosen profession. When he took a job in the legal department of ABC News, his bosses at the BEN Group didn't know who could fill his shoes. Who knows enough about Einstein to post ten times a day?

"Call Benyamin Cohen," Tony told them.

Flying from West Virginia to Israel is an incredibly long commute to meet my new coworkers. Especially considering I normally just roll out of bed and start tweeting Einstein quotes. But I was not going to miss an opportunity to introduce myself to this ragtag team of full-time Einstein enthusiasts. I needed to make a good first impression. The last thing I wanted to do is be that guy who spills coffee on the original handwritten general theory of relativity. Yes, they have it here. They also have Einstein's Nobel Prize, but I doubt it's lying around in the break room.

The archives are not in a state-of-the-art facility or even their own building. They reside below the dean's office, on the second floor of a classroom building. If you're walking down the hall, you'd be forgiven for walking right past the nondescript brown door that leads to a veritable treasure trove of Einstein history. But if you do knock and someone lets you in, you'll be greeted by the world's leading Einstein expert. He's a big man with broad shoulders and an impressive white beard, and he looks a little bit like a member of the band ZZ Top or Santa Claus. Except instead of a festive Christmas hat, he's wearing a skullcap the size of an upside-down cereal bowl. And forget the red robe; this guy is sporting a black silk frock coat typical of ultra-Orthodox Jews. He seems to have just walked out of a nineteenth century shtetl. Which may not be the worst thing, considering Einstein's upbringing in eastern Europe.

Roni Grosz—who, when you get a closer look, is most likely not Santa Claus—was raised in Austria and speaks fluent German, a prerequisite for his role here as chief curator, since many of the documents were written by Einstein in his mother tongue. Grosz took the job here in 2004 on a lark. "Growing up, I had no knowledge of Albert Einstein except that he was famous," he told me as we settle into his office. In the corner, I spotted a rather large safe, perhaps the final resting place of the original copy of Einstein's theory of relativity. I wanted to ask if what I thought was in there was actually in there and could I see it and take a selfie with it but, like I said, I was still trying to make a good first impression.

Like Einstein himself, Grosz spoke softly; you have to lean in just to hear him talk. He's a deep thinker and has a keen attention to detail, skills which serve him well here. He has become, for example, an expert in Einstein's signature—the way the "A" in "Albert" peaks like Everest and how the "E" in "Einstein" looks like a musical note. As it happens, people are always discovering new Einstein documents—some are authentic, while others have less-than-stellar provenance. Perhaps you're cleaning out your grandfather's attic and you stumble across a letter written to your family from the beloved scientist. Or you're surfing the web and spot an Einstein note for sale at an online auction. The first thing you'll want to do is email a photo of it to Grosz. Sometimes, he can take a quick gander on his computer and determine if it's a fake. Other times, he'll send it to the laser printer in his office and stand over it with a magnifying glass.

I asked him why he thinks Einstein's popularity has stood the test of time.

"It puzzles me as it puzzles you," he replied, in almost Yoda-like fashion. "But think about it. Practically no one that will tell you that Einstein was the greatest scientist understands his science. What are you appreciating? If you like a rock star, you say, 'I like his music.' You don't have to be a musician to appreciate it. And that's great. But here, if you don't understand the physics, what are you appreciating?" He trailed off as he answered his ringing phone. He muttered a few sentences in Hebrew and hung up.

"It reminds me of that time Charlie Chaplin invited Einstein to a premiere of one of his movies, *City Lights*," Grosz continued. "And they were sitting together, and people were applauding them. So, Einstein said to Chaplin, 'They're applauding you, not me! Because you're the star of this film.' And Charlie Chaplin told him, 'They cheer me because they all understand me, and they cheer you because no one understands you.'"

Grosz believes, like many do, that Einstein was the first icon of pop culture. "Although pop culture only developed after the Second World War, he was a precursor," he explained. "Before that, celebrities were members of the royal family. That's who everybody would recognize and that's who they would name their kids after."

Einstein was a global citizen, and likely the most famous American in the 1940s and 1950s. "Interestingly," Grosz explained, "all the others that followed after him, whether it was the Beatles or Marilyn Monroe, no matter how popular and how crazy people were at the time, with the decades, they fade away." I'm getting a little whiplash listening to a Hasidic Jew talk about Marilyn Monroe. But then he brought up the king of rock 'n' roll. "If you've seen films showing Elvis Presley concerts, people were really going nuts in ways unheard of before. But today, nobody's going crazy about Elvis Presley. They'll still say, 'Yeah, he was a good musician, and he was important.' But the emotional aspect is lost. But Einstein, no one of the young generation has ever seen him in their lifetime. And still, they're all crazy about him. It's a phenomenon."

Einstein's increasing popularity is on full display just outside of Grosz's office, in the entryway to the archives, where it seems every spare inch of space is taken up by some sort of Einstein knickknack: bobbleheads, puppets, paperweights, calendars, postcards, and T-shirts. And piles of photos you've never seen before: Einstein as a five-year-old boy, Einstein working on the theory of relativity, Einstein at home in Princeton. Every wall has a different Einstein poster. On one shelf, I spot Einstein's travel diaries. There's even an Einstein robot.

I'm like a kid in a candy store. But it's what's in the conference room that really excites me. That's where you'll find lining the walls more than one thousand books from Einstein's personal library. A sheet of glass protects the front of the bookcases, but Grosz slides open one section so I could touch the very books that Einstein touched. I was full of emotions.

It's an interesting mix of books. There were classics like *Gulliver's Travels*, *Vanity Fair*, a set of books by Leo Tolstoy and the complete works of Dostoyevsky. But there were also more obscure tomes that gave a slightly new insight into Einstein's vast interests: *Wonderful Australia in Pictures*, *The Conquest of Everest*, and a memoir by an Indian writer called *Prison and Chocolate Cake*. Not all of the books belonged to Albert: The large tome of *Grimm's Fairytales* was, I'm told, read by his stepdaughter Margot, who also lived in the house. A copy of *1984* by George Orwell had a handwritten note from the publisher asking Einstein to review the book. Another book had a burn inside, the result of ash dropping from Einstein's pipe.

Neill McManus, a charming man from the United Kingdom, is a paper conservator and is tasked with preserving these historical aspects at the archive. He brought me over to his workshop and we put on white gloves while he showed me original pages containing Einstein's math scribblings. One of them had a coffee cup stain on it. The pages literally have Einstein's fingerprints on them. McManus is careful to handle the original in an environment with appropriate temperature, humidity, and lighting. The originals are stored within sheets of polyester and sealed around the edges. He and the others here have scanned and digitized each page so researchers can search through the documents via the archives' online database.

There are eighty thousand documents—unfinished math equations, drafts of lectures, not to mention the many letters Einstein received during his lifetime. From dignitaries like President Roosevelt and Sigmund Freud to questions sent to him by children from around the world. Like a young girl named Monique who asked when the world would end. Einstein's reply? "Wait and see!" Another kid wrote to Einstein to tell him to cut his hair. Others were simply fan mail. A Manhattan plumber wrote in asking if Einstein would be his pipe partner. He wrote: "I am ready to change the name of my firm to Einstein and Stanley Plumbing Co."

Einstein responded to many of the letters, which are now in the hands of the descendants to whom he wrote. But those documents are not lost to the archive. Helen Dukas, Einstein's longtime secretary, did something very clever: She made a carbon copy of every letter they sent out. "This is extremely valuable," Grosz told me. "Because it gives us the back-and-forth of the correspondence, and this is what it's all about here—to build this mosaic or to weave this tapestry or whatever metaphor you want to use."

New documents are constantly being sent to the archives,

continuously expanding its library of content. On one of my subsequent visits to the archives, there was a press conference unveiling 110 new pages that had been donated by a collector in North Carolina.

Among the new items was what's known to Einstein-philes as "The Elusive Page 3," a document that had not been seen since 1930. It was then that the Nobel Prize–winning physicist published an article about his unified field theory with an appendix attached. Page 3 of that appendix had gone missing and was thought lost until now. "This is a rare find," exclaimed Grosz.

Overseeing the scientific side of the archives is Hanoch Gutfreund, a theoretical physicist like Einstein. He reminds me a little of Einstein, with his grandfatherly looks and thick accent. Gutfreund is the textbook definition of a company man: He's been at Hebrew University for over half a century. He began there as a freshman in college, got his PhD, became a physics professor, and ultimately became president of the entire school. Since 1997, he's been the academic director of the Einstein Archives, what he called "a cultural treasure of immense importance to mankind."

Gutfreund is in his late eighties, but he has the energy of someone much younger. He still comes to work every day; that is, when he's not jetting around the globe giving a lecture about Einstein, like at the conference in Paris or the opening of an Einstein exhibit in Taiwan. I tried scheduling a time to chat and it kept getting postponed: a phone call here, a meeting there. At one point, I saw him down the hallway and decided to chase after him. When I followed him along as we went from one building to another, he appeared to be speed-walking. And he had a backpack on, and his hands were full of papers. (I later learned he regularly runs 10K races.)

Gutfreund has a unique vantage point (ten feet ahead of me and in way better shape), as the scientific mind overseeing Einstein's enduring legacy. "The memory of Einstein refuses to recede into history," he told me. "The memory of great men and women usually fades away after their death. But, if anything, one can say that the interest in Einstein increases."

He admitted that this fascination applies even to him, someone whom you think already knows everything there is to know about Einstein. "The equations of the general theory of relativity are so complicated that they still hold secrets and phenomena which we will keep discovering. I, myself, discover something every day." I asked him what he learned yesterday. He told me he unearthed a letter Einstein wrote in the 1920s advising a German company that had a patent dispute with an American company. Apparently, they both made something called a "gyro-compass." But, Gutfreund said, his favorite document here at the archives is the original pages of the theory of relativity. "It's the magna carta of physics," he said with a quick smile, before dashing off to his next meeting.

The den mother of the archives is Chaya Becker, a sweet Canadian who takes care of everyone. A later visit I took to the archives coincided with Einstein's birthday. She brought in a cake, and we walked around the campus finding people to share it with. On the university quad, there's an eight-foot bronze statue of Einstein, and a class of high school kids was touring the grounds that day. I asked them to stand in front of the statue and sing happy birthday to the Nobel Prize winner. They happily obliged. We served them cake. And I livestreamed the whole thing on Facebook to Einstein's millions of followers.

If you have a random question about Einstein's life, reach out to Chaya, and she'll get you the information you need. I once asked her what Hebrew name Einstein's parents gave him at his circumcision. It took her a while, but she tracked it down, sending me an email with the subject line: "Mystery solved!" She had found an old handwritten registration document from the synagogue in Einstein's hometown of Ulm, Germany, on the banks of the Danube River. There it was, on the seventh line, scrawled there by a Rabbi Strauss: Albert's Hebrew name was the biblical Abraham.

On my second day in Israel, Chaya picked me up in her car and we drove to the town of Ra'anana, about a sixty-mile drive northwest of Jerusalem. She wanted me to meet an artist who has a pretty bizarre job at the Einstein Archives.

Enter the sprawling garden of Ido Agassi's backyard and into his studio and you're walking back in time. A poster advertising the 1957 Ford Thunderbird greets visitors over the main doorway. The furniture inside is a mix of farmhouse chic and midcentury modern. A construction worker's dusty lunchbox, for seemingly no apparent reason, hangs on a hook on one wall. But the main attraction is the vintage hand-cranked printing presses and other old machinery that was used by bookbinders in Einstein's era. Open a drawer and you'll see dozens of versions of different typefaces. In one, I find large block letters the size of my palm; in another, small six-point type. This is how you chose a font before computers.

Agassi, in his early forties, started a small press called Even Hoshen—the Hebrew phrase found in the Old Testament for the special stones that adorned the breastplate of the high priest—with his father, Uzi, who used to work in an art gallery in Tel Aviv. But this was not your typical publishing house. They only produce limited editions, and each book is, itself, a work of art. Agassi looks the part of a hipster bookbinder. He's wearing gray jeans and a black turtleneck with a bright red V-neck sweater over it and resembles a young, bearded Robert Redford.

He seems like he has an old soul, like he doesn't belong in the modern era. The machines he works with are twice as old as he is and use methods that are no longer required in the age of Amazon. He lays down each letter one at a time and you feel like this is perhaps how Gutenberg did it; indeed, there's not a computer in sight. For Agassi, it's a labor of love. "It's part of the art, the setting of the type, the manual labor in creating the book," he explained. "That's part of the beauty."

He handed me a pocket-sized book he made, which is in the shape of a slingshot. Inside, it tells the story of David and Goliath. Nestled into the binding is a tiny pebble. Agassi told me he went to the valley where the Biblical battle took place, about an hour drive from where we are, and sourced the rocks from there. Only thirty copies of his David and Goliath book exist.

Agassi is a craftsman and takes immense pride in his work. He calls himself a book artist—from the typography to paper selection and the type of binding to use, all of which is somehow related to the title or topic of the project. He worked with a local writer on a book of poems about ballet. In place of an actual binding is a pink ribbon in the shape of tutu skirts. It seems like a lost art, but Agassi and others like him meet up annually at what's known as the Codex Book Fair.

Agassi's place on the Einstein team is two-fold: First, he maintains the books from Einstein's personal library. Through the decades, the binding on that copy of *Gulliver's Travels* has weakened, for example, and he is tasked with repairing and strengthening it. But fixing the books is not always the solution. Einstein's dog, a fox terrier named Chico, liked to chew things— including some of the books on the shelf in Einstein's home. For Agassi, it provoked an existential question: To fix or not to fix? "We had the dilemma: should we fix and preserve the original binding or leave it as is?" he recalled of the bite marks. "We decided this is part of history, and we left it as is."

The second, and perhaps more fascinating, part of his job is making exact replicas of Einstein's forty-six-page theory of general relativity. The famous scientific document will sometimes go on display at museums in cities across the globe. But because the pages are more than one hundred years old, much care is taken to ensure their safety. (They have their own insurance policy.) Moreover, humidity extremes can potentially degrade the papers: low humidity can cause the papers to become dry and brittle while extremely high humidity can lead to mold.

So, Agassi was hired to make a perfect replica. He uses the same paper as Einstein and mimics everything from the original—including copying where Einstein crossed out mistakes and corrected himself. After one page has been out for ninety days at an exhibit, the curators often swap in one of Agassi's versions to minimize the light exposure to the original. This helps maintain the integrity of the pages. Museum goers are alerted ahead of time that some of the pages are exact replicas of the original. In 2016, which marked a century from the time the theory was published, Agassi made a limited edition of one hundred replicas of the theory of relativity for collectors of Einstein memorabilia.

"Touching his work and preserving it," he said, "it's a great honor."

I often wonder what Einstein would think of the team working at the archives, a group that now includes me. Sure, he would understand the work of Roni, Neill, Hanoch, Chaya, and Ido. They are ensuring that the legacy of his work is preserved. There is purpose to what they do: After all, academics who come here and sift through his unpublished manuscripts may, in fact, one day glean some new secret about the universe. But what about me? The guy who posts daily to Einstein's millions of fans on social media, something that did not exist in his era? I'm just giving people who check Instagram each morning a fun photo of the genius to start their day.

"Einstein was not particularly concerned what the masses

thought of him," Matthew Stanley, a professor of the history of science who wrote a book about Einstein, told me when I called him up. "The idea of going out of his way to express himself to the masses doesn't seem right. But he was happy to have his ideas presented. So, every afternoon at tea, if they would've said to him, 'Professor Einstein, you know, this happened this morning. What do you think about that?' He would have been up for that. So, I could totally imagine him having a social media coordinator."

My heart, it skipped a beat.

EINSTEIN AND
THE NEXT GENERATION

"I never think of the future.
It comes soon enough."

—ALBERT EINSTEIN

There are a handful of things I can recall from being a ten-year-old kid: My first crush, that one time I solved a Rubik's Cube, and the theme song to *Inspector Gadget*. For any kid like me who grew up on a steady diet of 1980s cartoons, we can still recall those ridiculously hummable musical notes that began each episode. While my siblings and I spent countless afternoons watching the dimwitted detective, it turns out that the guy who composed that inimitable theme song didn't spend too much time thinking about it. Music producer Shuki Levy composed the one-minute ditty while in the car driving to the studio.

Lest you think Levy was merely lucky, stopping at a red light and having an aha-moment that would net him fame and fortune, I should let you know that he is the exact opposite of a one-hit wonder. He also wrote the opening music to *He-Man*, *She-Ra*, *Dragon Quest*, and more than one hundred other TV theme songs. Think of any Reagan-era animated show and Levy was likely its composer. Not to mention, he's the co-creator of the *Power Rangers*, the iconic superhero franchise. That show has been on TV and in movies, in some form or another, for a quarter century.

Levy is less involved in kids' entertainment these days, choosing instead to spend time at his foundation, which he runs with his wife Tori, a popular food blogger and cookbook author. The nonprofit's goal is to support educational initiatives for children. Levy gave me and my friends *Inspector Gadget*, but he's hoping to leave a more lasting legacy for future generations. That's why Levy's foundation is helping build a new, multimillion-dollar home for the place where I work, the Albert Einstein Archives.

Levy travels back and forth often from his home in Los Angeles to his native Israel, and it was on a trip to Jerusalem where my colleagues—Roni, Chaya, Hanoch, and the rest of the gang—showed him around the current Einstein Archives. There was a need to help them find a bigger space, not only to house Einstein's eighty thousand documents, but also to increase the public awareness of Einstein's many achievements. To use the parlance of a fan of Levy's work, it was music to my ears.

The Einstein House as it will be known (Einstein personally never liked the term "museum") will be built on the site of an abandoned planetarium on the campus of Hebrew University. It will be two stories tall and include a visitor's center, a research area, conference rooms, a hall for events, and space for rotating exhibitions. The administrative employees of the archives will also move into the building, so this is where my office will be when I visit Israel.

Helping Levy foot the bill for the multimillion-dollar renovation is another creative, Jose Mugrabi, the world's leading collector of Andy Warhol paintings. (He has eight hundred, in case you were wondering.) Levy credits Einstein with being a source of inspiration in his Hollywood career. "To hear Einstein say things like, 'Imagination is more important than knowledge,' really connected with me strongly," Levy said, "because everything I've done in my life until now, it's where I am in my imagination."

When complete, the planetarium-turned-museum will be the largest repository of Einstein material in the entire world. It is just one of many projects aimed at spreading the gospel of Einstein to a new generation. Ellen DeGeneres is producing a reality show called *Finding Einstein*, which is scouring the globe in search of kids and young adults who think outside the box. "We're looking to find the great thinkers of the world in every genre—whether it's art or music or science or math," Doron Ofir, who is shepherding the casting process, told me. "We're looking for someone with a new way of thinking. Kids who can potentially be the keystone of changing humanity and the way it thinks."

In the past two decades, Ofir, a former nightclub doorman, has helped cast more than two hundred reality series that have aired on just about every major network. Ofir is the guy who cast the wildly popular *Jersey Shore* show on MTV. So, yes, you have him to thank for introducing the world to J-Wow and Snooki. He calls himself a "collector of characters."

So, he knows a little something about being a talent scout. "I'm excited about this project first and foremost because I feel like it's one that can actually change the world," he said. Ofir said he is looking for more than a kid with a cool science project. He's looking for the ones who are, like Einstein was described in that famous commercial for Apple computers, the misfits and the rebels. He's not looking for the polished pageant kids. "If you want to be on TV consistently," he said, "and you're forever trying to get on a reality show, you're probably not what I'm looking for." He cited the young climate activist Greta Thunberg as a model for the type of future Einstein they're looking to find. "It's a time of inspiration," Ofir said. "We want to inspire these kids to inspire us."

Einstein was so much more than merely a physicist. In many regards, his fame and symbolism as the world's favorite genius far outshined the theories of light and gravity and space-time that defined his early career. It is why so many children of his era felt a connection to him, the woolly-headed sage, who gave them the time of day. Throughout his life, Einstein corresponded with hundreds of children from across the globe: America, England, Japan, and South Africa, just to name a few of the return addresses that showed up on his Princeton doorstep. Some would ask questions—*Do scientists pray? What holds the sun and planets in space? How does color get into a bird's feather?*—and he'd write back with a thoughtful, and often witty, reply. With others, he developed an ongoing correspondence with them. "He respected children and liked their curiosity and fresh approach to life," Einstein's granddaughter, Evelyn, recalled. "I felt welcome in his presence because he made me feel very much at ease. Even when I was only five, he never talked down at me or intimidated me. I was astonished."

The Guinness World Record for the largest-ever gathering of Albert Einstein lookalikes took place in 2015 in California. Everyone donned bushy wigs and fake mustaches and smiled for the camera. Those assembled were not physicists. Heck, they weren't even adults. They were 304 elementary school students. Not to be outdone, in 2017, 404 students from an elementary school in Toronto broke the record. The current record holders are 541 students at a Catholic school in Australia. Kids who were born long after Einstein passed away are still being inspired by the genius.

Like many good ideas, this one was hatched after an evening of drinking. In 2010, three young Israelis—Yonatan Winetraub, Yariv Bash, and Kfir Damari—were at a bar in Holon, a small town just south of Tel Aviv. They made for an intriguing ensemble: a space engineer, a cybersecurity expert, and a drone maker. (Bash would later launch a company that builds drones that—I kid you not—deliver hamburgers to golfers.) As the night wore on, they devised an audacious plan to build a spacecraft that could land on the moon. You know, typical conversation after a few beers. "As the alcohol level in our blood rose, we got more and more determined to do this," Winetraub recalled when I spoke with him. "And it never faded away."

That booze-infused plan, written on the back of a napkin,

would eventually come to fruition, and their lunar aspirations couldn't have arrived at a more propitious time. For years, NASA had been moving away from moon missions, opting instead for routine trips to the International Space Station as well as hyping a glitzier journey to Mars. Google stepped in to fill the void. Back in 2007, the Silicon Valley tech giant established the Google X Prize. It offered $20 million to the first team to land an unmanned rover on the moon and send back high-definition video. Groups from all over the world signed up, hoping to take part in this modern-day space race.

The three Israelis—who called their team SpaceIL—excelled in the competition, quickly separating themselves from the pack by becoming one of the top three teams. As the contest waged on, Google kept pushing back the deadline to allow more time for the teams to work. In early 2018, with no team able to meet the latest deadline, Google withdrew the prize money. But SpaceIL was so close to finishing the project, only months away from completion, that they decided to keep going.

The ship was finally completed in late 2018. A dedication ceremony that December christened the spacecraft the *Beresheet*, Hebrew for "Genesis." In January 2019, it was shipped from Israel to the United States, where it took up residence at Cape Canaveral to undergo final tests before the launch. If the mission proved successful, the tiny nation of Israel—it's practically the size of New Jersey—would become only the fourth country to ever successfully land on the moon, after the United States, Russia, and China. The mission had all the ingredients of a made-for-Hollywood movie. The world watched—on February

22, 2019—as the unmanned ship lifted off from the Kennedy Space Center in Florida. The entire mission, Winetraub told me, was inspired by Albert Einstein.

When Einstein first spoke about his theory of relativity, he was brushed off by many in the scientific community. After all, he was upending the entirety of physics and, with it, our understanding of the universe. Likewise, the SpaceIL team was navigating new territory by attempting to do something nobody else had done before—become the first privately funded ship to land on the moon. All previous spacecrafts were paid for by their respective governments. To accomplish this, the three Israeli entrepreneurs had to develop a way to create a ship on a shoestring budget. That meant compromises had to be made: redundant backup systems, which are a standard safety precaution in space travel, were tossed out in favor of crossed fingers and wishes for good luck. They couldn't afford to design parts to their specifications, so they found some that looked kind of similar in an online catalog. "At NASA, we would never build a spacecraft with off-the-shelf stuff," said Dr. Philip Metzger, a planetary physicist, when he heard about the team's plans. SpaceIL cut so many corners, they would only be able to communicate with the ship when it was facing a certain direction. Moreover, the machine's engine could only be turned on for brief spurts at a time. For the most part, the *Beresheet* ship relied on good old-fashioned gravity to make the 239,000-mile journey from Earth to the moon.

The lightweight *Beresheet* was about the size of a coffee table and, if it proved successful, would be the smallest spacecraft to ever make a lunar landing. They cut costs further by hitching a ride on the back of an existing rocket from Elon Musk's SpaceX that was already at the launch pad. Think of it like an Uber carpool ride but, you know, to outer space. "It's less of a piggyback, and more like a rideshare," Winetraub explained. The SpaceX rocket was carrying loads for several missions and was going to drop each one off at a different point in space. A direct trip to the moon would normally only take days. But because SpaceIL took this budget route—getting dropped off and then basically having to float the rest of the way—it would take about two months to arrive at the moon.

Most moon missions cost billions of dollars. Their low-cost rookie moonshot was built for $100 million—which they raised through private donations. "This is an investment in science," Sylvan Adams, a philanthropist who helped fund the project, told me when we spoke in Jerusalem. "This is an investment in inspiring our youth to think beyond the atmosphere and beyond our planet."

The scrappy scientists persevered, like Einstein, armed with an underdog ethos. Landing the *Beresheet* on the moon was more than just about bragging rights for Israel. The mere fact that they got it to the launch pad and into orbit was success enough. Aboard the ship was a time capsule containing three DVDs with hundreds of digital files. On one was the so-called Lunar Library, a thirty-million-page archive of human history and civilization. In case anything happened to our planet, SpaceIL was hoping that this "civilization backup" would remain on the moon as a record of our time on Earth. On another disk were files of drawings—of the moon, of space, and of the universe—all made by children. Winetraub, who had interned at NASA, wanted to show the youth back in his country that they too could touch the cosmos. It was a lesson he had learned from Einstein. "He wasn't just a physicist. He was also doing a lot of outreach," Winetraub said, adding that Einstein rarely spoke in technical terms. "He was all about giving the story of what his complicated physics was to the public. And I think, to some extent, it's the same thing we're trying to do—on a smaller scale obviously. This is not general relativity." He paused, before adding: "It's all relative."

There were no astronauts on board the ship, and all of its navigation was controlled manually by the team down on Earth. They used a laser guidance system—a technology built upon Einstein's photoelectric effect—to figure out the right path. "Einstein was an excellent physicist and an excellent communicator. We're trying to do some of that ourselves," Winetraub said. "It's a really complicated engineering problem, but we want to be able to communicate that so that kids would one day be able to build their own spaceship." It's the impact his moon mission is making on the next generation that is having the biggest effect on Winetraub. "The thing that touched me the most was that I was waiting for a friend just outside a coffee shop in Israel. And there were some kids that were playing around. One of them asked me if I was one of the people from SpaceIL. The kids recognized me, and they wanted to take a selfie with me."

Einstein was used to having the paparazzi tracking his every move. "The fact that the engineers are now, to some extent, becoming celebrities is quite remarkable," Winetraub said.

"Kids want to be space engineers when they grow up, and that's really touching. Because we did this mission to inspire kids. I feel that this is one of the most beautiful things of this mission." The *Beresheet*'s journey in space coincided with the festival of Purim, a sort of Jewish Halloween, when kids dress up in costumes. During that spring, as the ship slowly floated toward the moon, there were Purim parades in Israel where kids dressed up as astronauts and strolled down the street alongside a replica of the *Beresheet*. "A year ago, everyone was dressed up as Wonder Woman," Winetraub said with a laugh. "So, it's quite a contrast."

After spending six weeks flying closer and closer to the moon, the ship prepared for landing on April 11, 2019. In Israel, Winetraub and dozens of engineers huddled around the monitors at the command center. Israeli Prime Minister Benjamin Netanyahu was there to congratulate them. There were viewing parties across the globe as millions watched with bated breath, hoping to witness another moment as historic as the lunar landing, exactly fifty years earlier, of Buzz Aldrin and Neil Armstrong. Winetraub gave me access to the feed, and I streamed it live for public viewing on Albert Einstein's Facebook page.

"We have passed the point of no return," came the voice from the control room, as the ship began its final descent. "We are in the landing process." Winetraub and his colleagues had retrofitted *Beresheet* to land by itself using nothing more than a homemade flight simulator. It made its way closer and closer as a camera on an outstretched metal arm took pictures of the journey. Thirteen miles above the lunar surface, the *Beresheet* took a selfie with the moon in the background and sent it back to Earth. It would be its final act. The ship started speeding up instead of slowing down. Seconds later, it crashed into the surface of the moon.

"Condolences to the *Beresheet* lander," an eighty-nine-year-old Buzz Aldrin posted on Twitter. "It couldn't quite stick the landing. Never lose hope—your hard work, teamwork, and innovation is inspiring to us all." Nine-year-old Ori Solow was one of those Israeli kids who dressed up in a homemade *Beresheet* costume made out of cardboard and duct tape for Purim. He stayed up late into the night to watch the moon landing. I spoke with his mom, Atara, afterward. "It's particularly meaningful for me to encourage him to dare, think creatively, innovate, and seek to make an impact," she said. "Who knows? Maybe one day he will be developing the next space technology." She said the fact that the ship didn't make it all the way was perhaps more meaningful. "What a strong lesson for our children. There's no need for disappointment. We will get there eventually. It's just a great and important accomplishment that hasn't yet been completed."

The mission captured everyone's imagination and ushered in a new era of space flight. SpaceIL is already planning a *Beresheet 2*.

We began this book with a story about one of the largest telescopes on Earth, located in the middle of an empty field in West Virginia. And a Harvard astronomer named Avi Loeb, looking for aliens on a quest inspired by Einstein. And we're going to wrap up with another telescope. This one is the most powerful telescope ever built, and it's not located in the

United States or even on planet Earth. The James Webb Space Telescope, as the name implies, is in outer space. And it also only works because of Albert Einstein.

The James Webb Space Telescope, at a cost of $10 billion, was one of the most complex missions NASA had ever attempted. So many things could have gone wrong: 344 things to be exact. If any of those single points failed, the entire mission would fail. To put this into some perspective, a mission to land on Mars has between seventy and eighty-five potential areas that can fail. In other words: a whole lot can go disastrously wrong. Compounding the problem: If something broke on the Hubble telescope, the precursor to the Webb, it was close enough to Earth that NASA could send astronauts to fix it. There's no such luxury with the Webb—it's floating one million miles away. Three decades of work by more than twenty thousand scientists, engineers, technicians, and designers could literally go up in smoke in an instant.

There were no backups. NASA had one shot to get it right.

The James Webb Space Telescope launched on Christmas day in 2021 aboard a rocket from a tropical rain forest in French Guiana in South America. (Being near the equator allows you to take advantage of the rotation of the Earth and add velocity to your liftoff.) Once the rocket arrived in space, the team at NASA watched as Webb began to unpack itself, and they breathed a collective sigh of relief when all 344 things worked exactly as planned.

When the first photos taken by the Webb were revealed at a press conference at the White House in 2022, scientists wept. The photos, allowing us to peer deeper into the universe than ever before, were utterly breathtaking. One reporter described looking at the photos in slack-jawed wonder. You should stop reading this book, just for a moment, and Google the images. I called up Loeb to get his thoughts. "The luminous cores of galaxies are like fish swimming in a container filled with transparent water, bound together by gravity—which serves as the 'aquarium' walls," he said, quite eloquently.

And we have Einstein to thank for the clarity of these images. That's because of a concept called "gravitational lensing," first identified by Einstein in 1912 and later published in a paper in 1936, which allows us to see galaxies arc and bend. It's basically the difference between seeing a flat image of the universe and one that's in 3D. Gravitational lensing provides depth to the photos, giving us greater detail of the foreground and background and the gasses in between. Distant galaxies appear small, much the same way mountains would appear tiny in a photo you take while standing very far away. In these photos from space, the lensing creates a distinctive circle around the galaxy. Scientists call them "Einstein rings."

Loeb said that many astronomers are looking to the Webb to confirm scientific facts that we already know, but he is less interested in that. "What I'm really, really looking forward to are the surprises," said Loeb, who previously served on the Webb's advisory board and wrote an early textbook about the Webb's potential. "The thrill of doing science is that it can be exciting, especially when nature is trying to educate us and teach us something new." He drew a comparison to Einstein, whose special theory of relativity was not expected by the mainstream scientific community. But the study of physics was propelled by Einstein's unexpected discoveries. "Don't expect what you will find looking at the sky, and you might find something that you don't expect. And that would be a big thrill."

On the Monday morning of Albert Einstein's death in 1955, when a certain pathologist was busy sawing out the brain of the world's favorite genius, *Life* magazine dispatched photographer Ralph Morse to Princeton. "I headed to the hospital first," Morse recalled years later, "but it was chaos—journalists, photographers, onlookers. So, I headed over to Einstein's office. On the way, I stopped and bought a case of scotch. I knew people might be reluctant to talk but that they're usually happy to accept a bottle of booze, rather than money, in exchange for their help. Anyway, I get to the building, find the superintendent, offer him a fifth of scotch and like that, he opens up the office."

Once inside, Morse was alone with Einstein's desk. It was covered in a mess of books and stacks of papers. "The empty chair by the formula-filled blackboard," the magazine wrote alongside Morse's iconic photo, "looked as if the scholar who usually sat in it had merely stepped away, perhaps to gaze reflectively at the meadow, which rolls past the Princeton Institute for Advanced Study. But the chair would not again be filled." A pipe rested on a notebook that was propped open.

Up until the very end, Einstein had been working on his "theory of everything." It had frustrated him for years, and it was his quixotic attempt to merge his theory of relativity—which looks at phenomena like black holes in the vastness of space—and quantum theory—which explores the most minuscule atomic matter. He tried for decades, ultimately unsuccessfully, to comprehend how those two theories could coexist. But that search for Einstein's theory of everything now continues. "Today, many of the world's top physicists are embarking on this cosmic quest, whose far-reaching reverberations span our understanding of reality and the meaning of existence," wrote Michio Kaku, a noted physics professor. "It would be the crowning achievement of thousands of years of scientific investigation, since ancient civilizations also wondered how the universe was created and what it is made of."

So where do we go from here? We live in a bifurcated world, to say the least. Countries are divided by those on the right and those on the left, with both sides refusing to connect on middle ground. Those who obsessively watch reality TV are likely not the same people who are attempting to build time machines. My life in West Virginia is markedly different from my brother's in New York City. And both of us have no comprehension what daily life is like in India. We all live comfortably in our own filter bubbles. And yet, somehow, we have all been beneficiaries of the Einstein effect.

A taxi driver in Paris is finding his way thanks to Einstein's GPS. A baker in Istanbul is winning an international competition for a cake in the shape of Einstein's head. A Harvard astronomer is searching for aliens, an Emmy-winning actor is helping rescue refugees, and a kid in Israel thinks he can build a next-generation rocket to space, all because of a near mythic creature who is more popular today than he was when he was alive. Young or old, Republican or Democrat, wealthy or poor—everyone feels some sort of connection to Albert Einstein. Some may be trying to photograph a black hole in outer space, while others are simply hanging a poster of him on their dorm room wall. Some are watching Einstein hunt criminals on TV, while others are using his theories to build the technologies of tomorrow.

Now, perhaps more than ever, we hunger for a figure that we can all coalesce around. Politicians are divisive and celebrities come and go, but Einstein and his transformative ideas have withstood the test of time. He has impacted multitudes. "His work," explained science writer Sam Kean, "strums the deepest chords of our imagination."

And so, with the weight of all that on my feeble shoulders, I will wake up tomorrow morning and help my neighbor milk her cows. I will collect the eggs from my chickens, including the one named for the world's favorite genius. I will pour myself a cup of coffee and log onto the internet, where tens of millions of fans look to me as the face of Albert Einstein. In those moments at the beginning of each day, when I stare at a blank screen and a blinking cursor, I will be guided by Einstein's principle of living a life that is "passionately curious." I will seek out discoveries and ask questions that don't yet have answers. I will compose a tweet or post a photo in the hopes that it inspires people to view the world in a new light.

And then I will hit Enter.

ACKNOWLEDGMENTS

"What I have to say about this book
can be found inside the book."

—ALBERT EINSTEIN, REPLYING TO A *NEW YORK
TIMES* REPORTER'S REQUEST FOR A COMMENT ABOUT
EINSTEIN'S BOOK, *THE EVOLUTION OF PHYSICS*

Writing a book about one of the smartest people of the past century made me feel quite dumb. So I owe a tremendous debt of gratitude to the folks who lifted me up and helped through this process.

First and foremost, my sincerest thanks to Tony Iliakostas, who originally played the role of Einstein on social media and graciously brought me into the fold. He deserves his own Nobel Prize. Thanks to Paul Kasko for encouraging me to write endlessly about Einstein. To Michael Paterniti for sparking my decades-long obsession with Einstein and to Carolyn Abraham for serving as an unparalleled guide during the writing of this book, including pointing me to where I could find the last physical remnants of Albert Einstein's brain.

Thanks to my agent, Jason Yarn, for believing in this project from the jump and offering up keen insights that turned my scattered ideas into a cogent thesis. To my editor, Erin McClary, who never wavered in her support of the project, even when I blew through two deadlines. And to Dominique Raccah, Madeleine Brown, and the rest of the Sourcebooks family for elevating my work.

To the group of early readers who were always available to take a first look at the chapters as I wrote and offered their feedback and words of encouragement: Micah Fierstein, Ayelet Polter, Chana Shapiro, my sister Chanie Kirschner, and my dad and stepmom, Herbert and Meryl Cohen.

To everyone at the Albert Einstein Archives who has welcomed me to the team: Ido Agassi, Chaya Becker, Roni Grosz, Hanoch Gutfreund, Neill McManus, and Avi Muller. Thanks to Liz Van Denburg Cohen for entrusting me to be the face of Einstein to millions.

To the coterie of Einsteins I had the pleasure to meet: Cayla, Daniel, Jeff, Laurie, Mara, Rebecca, Sarah, Stephen, and Ted. And to Karen Cortell Reisman who is the living embodiment of the Einstein name.

To all of those who graciously took the time to be interviewed for this book, including: Sylvan Adams, George Biddle, Jake Broder, Dawn Butler, the other Ben Cohen, David Ellenstein, Harald Geisler, Michael Gordin, Lauren Gussis, Lawrence Haber, Duffy Hudson, Daniel Kennefick, Michelle LaBar, Fred Lepore, Shuki Levy, Lia Li, Avi Loeb, Duncan Lorimer, Jim Lynch, Alec MacGillis, Ronald Mallett, Maura McLaughlin, Sean McWilliams, Jeff Newelt, Doron Ofir, Inci Orfanli-Erol, Mandy Patinkin, Lenny Pozner, Anna Pysana, John Reznikoff, Yahoo Serious, Matthew Stanley, Dr. X, Rebecca

Watson, Gladys Mae West, and Yonatan Winetraub. And thanks to Michael Geller, Stephen Kurczy, Carolyn Oglesby, Stanford Prescott, and the team at the International Rescue Committee.

To the scientists who helped explain Einstein to an Average Joe like me: you are true geniuses. Any mistakes in writing about theoretical physics, black holes, and quantum mechanics are, of course, my own.

To Josh Feingold, Sharon Feingold, Hallie Raab, Joey Raab, Chaim Saiman, and Yoel Spotts for being lifelong friends. To Scott Jones for being a great conversationalist, at any hour. To Lydell Owens for always asking how the writing was coming along. To John Temple for being an early sounding board. To Billy Ray and Jonny Umansky for being my go-to guys in Hollywood. To AJ Jacobs, who has been a champion of my writing career for eighteen years and continues to be a fount of inspiration—even though he took the title *It's All Relative* for his own book. Still, a true mensch.

To Jodi Rudoren who, like Einstein himself, found a way to harness time and then gifted it to me. Her daily editing of my work has made me a better writer. And to my other colleagues at the *Forward*, who are some of the smartest journalists I know: they each, in their own way, played a role in helping with this manuscript.

To my wife, Elizabeth, for allowing me to fulfill my *Funny Farm* fantasy of writing a book while living in the woods.

And to Albert Einstein, the world's favorite genius, for his boundless gifts.

BIBLIOGRAPHY

Abraham, Carolyn. *Possessing Genius: The Bizarre Odyssey of Einstein's Brain.* Toronto: Penguin Canada, 2002.

Calaprice, Alice. *The Ultimate Quotable Einstein.* Princeton, NJ: Princeton University Press, 2013.

Calle, Carlos I. *Einstein for Dummies.* Hoboken, NJ: Wiley, 2005.

Clegg, Brian. *How to Build a Time Machine: The Real Science of Time Travel.* New York: St. Martin's Griffin, 2013.

Einstein, Albert, Alice Calaprice, and Evelyn Einstein. *Dear Professor Einstein: Albert Einstein's Letters to and from Children.* Amherst, NY: Prometheus Books, 2002.

Friedman, Alan J., and Carol C. Donley. *Einstein as Myth and Muse.* Cambridge: Cambridge University Press, 1985.

Gott, J. Richard. *Time Travel in Einstein's Universe: The Physical Possibilities of Travel through Time.* Boston: Houghton Mifflin, 2001.

Herneck, Friedrich. *Einstein at Home.* Amherst: Prometheus, 2016.

Holton, Gerald, and Yehuda Elkana. *Albert Einstein, Historical and Cultural Perspectives: The Centennial Symposium in Jerusalem.* Princeton, NJ: Princeton University Press, 2016.

Isaacson, Walter. *Einstein: His Life and Universe.* New York: Simon & Schuster, 2007.

Jacobs, A.J. *The Puzzler: One Man's Quest to Solve the Most Baffling Puzzles Ever, from Crosswords to Jigsaws to the Meaning of Life.* New York: Crown, 2022.

Jerome, Fred. *The Einstein File: J. Edgar Hoover's Secret War against the World's Most Famous Scientist.* New York: St. Martin's Griffin, 2002.

Jerome, Fred, and Rodger Taylor. *Einstein on Race and Racism.* New Brunswick, NJ: Rutgers University Press, 2006.

Kennefick, Daniel J. *No Shadow of a Doubt: The 1919 Eclipse That Confirmed Einstein's Theory of Relativity.* Princeton University Press, 2019.

Kurczy, Stephen. *The Quiet Zone: Unraveling the Mystery of a Town Suspended in Silence.* New York: Dey Street Books, 2021.

Lepore, Frederick E. *Finding Einstein's Brain.* New Brunswick, NJ: Rutgers University Press, 2018.

Loeb, Avi. *Extraterrestrial: The First Sign of Intelligent Life Beyond Earth.* Boston: Houghton Mifflin Harcourt, 2021.

Mallett, Ronald L., and Bruce B. Henderson. *Time Traveler: A Scientist's Personal Mission to Make Time Travel a Reality.* New York: Thunder's Mouth Press, 2006.

Paterniti, Michael. *Driving Mr. Albert.* New York: Dial Press, 2000.

Stanley, Matthew. *Einstein's War: How Relativity Triumphed Amid the Vicious Nationalism of World War I.* New York: Dutton, 2019.

Williamson, Elizabeth. *Sandy Hook: An American Tragedy and the Battle for Truth.* New York: Dutton, 2022.

NOTES

EINSTEIN TWEETS

sent out a message to Ivanka: Einstein, Albert. "We Can Confirm That Albert Einstein Never Said This Quote. Here's a Worthy Purchase via @Princetonupress: https://T.co/fdgwko1qpz Https://T.co/Ewtuhskgvr." Twitter, July 24, 2017. https://twitter.com/AlbertEinstein/status/889474132106203136.

The Huffington Post's headline: Mazza, Ed. "Ivanka Trump's Attempt to Quote Albert Einstein Backfires Spectacularly." *HuffPost*, July 25, 2017. https://www.huffpost.com/entry/ivanka-trump-albert-einstein_n_59769c6ae4b0a8a40e81a994.

Newsweek declared: Sinclair, Harriet. "Ivanka Trump Misquoted Einstein and Now People Are Sharing Their Own Fake Quotes." *Newsweek*, July 23, 2017. https://www.newsweek.com/ivanka-trump-misquoted-einstein-and-internet-loves-it-640774.

"Pressmen roared up the plank": Abraham, Carolyn. *Possessing Genius: The Bizarre Odyssey of Einstein's Brain.* Toronto: Penguin Canada, 2002, 158.

dubbed Albert "the dope": Parkin, Simon. "Who Owns Einstein? The Battle for the World's Most Famous Face." *The Guardian*, May 17, 2022. https://www.theguardian.com/media/2022/may/17/who-owns-einstein-the-battle-for-the-worlds-most-famous-face.

When Einstein gave a speech: "4,000 Bewildered as Einstein Speaks." *New York Times*, June 17, 1930. https://www.nytimes.com/1930/06/17/archives/4000-bewildered-as-einstein-speaks-scientist-before-world-power.html.

Awe-inspiring articles: Berger, Jonah. Essay. In *Contagious: Why Things Catch On.* New York: Simon & Schuster, 2013, 93–124.

"The most beautiful emotion": Calaprice, Alice. *The Ultimate Quotable Einstein.* Princeton, NJ: Princeton University Press, 2013, 331.

an eco-friendly refrigerator: Panganiban, Roma. "Einstein's Design for a Fridge to Last 100 Years." *Mental Floss,* March 14, 2013. https://www.mentalfloss.com /article/49282/einsteins-design-fridge-last-100-years.

making all of us a lot dumber: Nichols, Thomas M. *The Death of Expertise: The Campaign against Established Knowledge and Why It Matters.* Oxford: Oxford University Press, 2018, Chapter 4.

Donald Trump famously cited himself: Collins, Eliza. "Trump: I Consult Myself on Foreign Policy." *Politico,* March 16, 2016. https://www.politico .com/blogs/2016-gop-primary-live-updates-and-results/2016/03/trump -foreign-policy-adviser-220853.

are being accused of writing fake news: Hayes, Christal. "Trump: Pulitzers Awarded to NYT, Washington Post Should Be Revoked for 'Fake' Russia Coverage." *USA Today,* March 30, 2019. https://www.usatoday.com/story /news/politics/2019/03/29/president-trump-pulitzer-new-york-times -washington-post/3316086002/.

a 2020 law passed in Ohio: Hancock, Laura. "Ohio Senate Unanimously Passes Student Religious Expression Bill." *Cleveland.com,* June 10, 2020. https://www .cleveland.com/open/2020/06/ohio-senate-unanimously-passes-student -religious-expression-bill.html.

against their religion: Hancock, Laura. "Ohio Lawmakers Clear Bill Critics Say Could Expand Religion in Public Schools." *Cleveland.com,* November 14, 2019. https://www.cleveland.com/open/2019/11/ohio-lawmakers-clear-bill -allowing-students-to-turn-in-inaccurate-work-in-name-of-religion-second-anti -science-bill-in-a-week.html.

"If you think intelligence is dangerous": Calaprice, Alice. *The Ultimate Quotable Einstein.* Princeton, NJ: Princeton University Press, 2013, 478.

Person of the Century: Golden, Frederic. "Albert Einstein." *Time,* December 31, 1999. http://content.time.com/time/magazine/article/0,9171,993017,00.html.

"It's like trying to drink from a firehose": Fred Lepore, Zoom interview with the author, June 5, 2021.

STEALING EINSTEIN'S BRAIN

"The last words of the intellectual giant": "Dr. Albert Einstein Dies in Sleep at 76; World Mourns Loss of Great Scientist." *New York Times,* April 19,

1955. https://timesmachine.nytimes.com/timesmachine/1955/04/19/93802835.pdf.

"a big blister": Lepore, Frederick E. *Finding Einstein's Brain.* New Brunswick, NJ: Rutgers University Press, 2018, 23.

"Let it burst": Abraham, Carolyn. *Possessing Genius: The Bizarre Odyssey of Einstein's Brain.* Toronto: Penguin Canada, 2002, 48.

Henry Abrams: "Einstein's Doctor Kept Scientist's Eyes." *UPI,* December 21, 1994. https://www.upi.com/Archives/1994/12/21/Einsteins-doctor-kept-scientists-eyes/3189787986000/.

Michael Jackson: Abraham, *Possessing Genius: The Bizarre Odyssey of Einstein's Brain,* 275.

"the wishes of the deceased": Ibid., 85.

ten small pine boxes: Ibid., 102.

slicing it into 30,953 wafers: Bentivoglio, Marina. "Cortical Structure and Mental Skills: Oskar Vogt and the Legacy of Lenin's Brain." *Brain Research Bulletin.* Elsevier, December 29, 1998. https://www.sciencedirect.com/science/article/abs/pii/S0361923098001245.

"multiplicity of his ideas": Abraham, *Possessing Genius: The Bizarre Odyssey of Einstein's Brain,* 30.

handed it over to a bumbling assistant: Gosline, Sheldon Lee. "'I Am a Fool': Dr. Henry Cattell's Private Confession about What Happened to Whitman's Brain." Iowa University Libraries. Accessed August 19, 2022. https://pubs.lib.uiowa.edu/wwqr/article/25903/galley/134271/view/.

An article in the *Journal of Sex Research*: Bierman, Stanley M. "The Peripatetic Posthumous Peregrination of Napoleon's Penis." Taylor & Francis. Accessed August 19, 2022. https://www.tandfonline.com/doi/abs/10.1080/00224499209551669.

sent four pieces in a mayonnaise jar: Carolyn Abraham, interview with the author, July 29, 2022.

"I want you to find Einstein's brain": Levy, Steven. "Yes, I Found Einstein's Brain." *Wired,* April 17, 2015. https://www.wired.com/2015/04/yes-i-found-einsteins-brain/.

"granted a rare peek into an organic crystal ball": Levy, Steven. "My Search for Einstein's Brain." *New Jersey Monthly,* August 1, 1978. https://njmonthly.com/articles/historic-jersey/the-search-for-einsteins-brain/.

"Good Morning America called": Abraham, *Possessing Genius: The Bizarre Odyssey of Einstein's Brain,* 174.

reminded him of Goldenberg's Peanut Chews: Levy, Steven. "Yes, I Found Einstein's Brain." *Wired*, April 17, 2015. https://www.wired.com/2015/04 /yes-i-found-einsteins-brain/.

"It reminded me of Jesus Christ": Abraham, *Possessing Genius: The Bizarre Odyssey of Einstein's Brain*, 264.

a group of rabbis: Ibid., 263.

"as an amulet for a necklace": Ibid., 314.

"You have a unique take on Einstein": Carolyn Abraham, interview with the author, November 9, 2021.

"I'd better stop at customs": Ibid., 283.

"This unusual brain anatomy": Witelson, Sandra, Debra Kigar, and Thomas Harvey. "The Exceptional Brain of Albert Einstein." *The Lancet*, June 19, 1999. http://www.bic.mni.mcgill.ca/users/elise/Alberts_brain.pdf.

"bigger where it counts" Abraham, *Possessing Genius: The Bizarre Odyssey of Einstein's Brain*, 322.

they built an iPad app: Johnson, Carla K. "Einstein App Boggles Mind." *New York Post*, September 26, 2012. https://nypost.com/2012/09/26/einstein-app -boggles-mind/.

"I didn't hear an ethereal choir": Fred Lepore, interview with the author, June 5, 2021.

first-of-its-kind paper: Men, Weiwei, Dean Falk, Tao Sun, Weibo Chen, Jianqi Li, Dazhi Yin, Lili Zang, and Mingxia Fan. "Corpus Callosum of Albert Einstein's Brain: Another Clue to His High Intelligence?" *Brain*. Oxford University Press, September 21, 2013. https://academic.oup.com/brain/article/137/4 /e268/365419.

"a Mozart from a Manson": Abraham, *Possessing Genius: The Bizarre Odyssey of Einstein's Brain*, 129.

a frequent guest on the *Late Show with David Letterman*: Strausbaugh, John. "A Curator's Tastes Were All Too Human." *New York Times*, October 11, 2005. https://www.nytimes.com/2005/10/11/arts/design/a-curators-tastes-were -all-too-human.html.

sixty thousand tourists annually: "Mutter Museum Gets Even Weirder." *Today Show*, April 12, 2005. https://www.today.com/popculture/mutter-museum-gets -even-weirder-wbna7477992.

"a quiet custodian": "A Beautiful Mind: Einstein's Brain Samples on Display at Mutter." *WHYY*, November 18, 2011. https://whyy.org/articles/a-beautiful -mind-einsteins-brain-samples-on-display-at-mutter/.

"This is Einstein's brain": Dr. X, interview with the author, February 11, 2022.

animals could live in Pleistocene Park: Mezrich, Ben. *Woolly: The True Story of the Quest to Revive One of History's Most Iconic Extinct Creatures.* New York: Atria Books, 2017.

FLYING, DRIVING, AND SURFING WITH EINSTEIN

has a nickname: Clark, Anders. "Cirrus SR22: The Plane with the Parachute." *Disciples of Flight,* July 5, 2017. https://disciplesofflight.com/cirrus-sr22/.

"I did not start out": Gladys Mae West, interview with the author, February 3, 2022.

"the regularity of Pluto's motion": "Mathematician Inducted into Space and Missiles Pioneers Hall of Fame." Air Force Space Command (Archived), December 7, 2018. https://www.afspc.af.mil/News/Article-Display/Article/1707464/mathematician-inducted-into-space-and-missiles-pioneers-hall-of-fame/.

"Without Einstein's theory": Dijkgraaf, Robbert. "The Uselessness of Useful Knowledge." *Quanta Magazine,* October 20, 2021. https://www.quantamagazine.org/science-has-entered-a-new-era-of-alchemy-good-20211020/.

India has since: Srivastava, Ishan. "How Kargil Spurred India to Design Own GPS—Times of India." *Times of India,* April 5, 2014. https://timesofindia.indiatimes.com/home/science/how-kargil-spurred-india-to-design-own-gps/articleshow/33254691.cms.

"I'm a doer": Mohdin, Aamna. "Gladys West: The Hidden Figure Who Helped Invent GPS." *The Guardian,* November 19, 2020. https://www.theguardian.com/society/2020/nov/19/gladys-west-the-hidden-figure-who-helped-invent-gps.

When BMW first introduced: Stober, Dan. "Professor Studies What Cars Can Learn from Drivers' Words." Stanford University, May 7, 2008. https://news.stanford.edu/news/2008/may7/cars-050708.html.

There were countless news reports: Cohan, Peter. "Apple Maps' Six Most Epic Fails." *Forbes,* October 9, 2012. https://www.forbes.com/sites/petercohan/2012/09/27/apple-maps-six-most-epic-fails/.

operated in more than 10,500 cities: "Uber Technologies, Inc." Sec.gov, December 31, 2021. https://www.sec.gov/ix?doc=%2FArchives%2Fedgar%2Fdata%2F1543151%2F000154315122000008%2Fuber-20211231.htm.

The Nebraska Republican: Wright, David. "Senator from Nebraska Moonlights as Uber Driver." CNN, November 14, 2016. https://www.cnn.com/2016/11/14/politics/ben-sasse-uber-driver.

Another is Elwood Edwards: Criss, Doug. "When the 'You've Got Mail'

Guy Is Your Uber Driver." CNN, November 7, 2016. https://www.cnn
.com/2016/11/07/us/elwood-edwards-uber-trnd.

that is Shaquille O'Neal: "Undercover Lyft." YouTube. Accessed August 20, 2022.
https://www.youtube.com/playlist?list=PL-04sKrMar6Nnjw-94V1zSjpg
EDahng1X.

"Your neurological system": Michelle LaBar, interview with the author, December
17, 2021.

"If we're the last people": Harris, Michael. *The End of Absence: Reclaiming What
We've Lost in a World of Constant Connection.* United States: HarperCollins
Canada, 2014, 16.

Comcast landed: Gibson, Kate. "America's Most Hated Companies." CBS News,
February 1, 2018. https://www.cbsnews.com/news/americas-most-hated
-companies/.

Our story made the six o'clock news: "Fighting for Connectivity." *WDTV,*
June 24, 2020. https://www.facebook.com/benyamincohen/videos/101583793
73698622.

Some backyard astronomers: Odenwald, Sten. "Starlink Satellite Streaks: How Big
of a Problem Are They?" *Astronomy.com,* March 25, 2022. https://astronomy
.com/news/2022/03/starlink-satellite-streaks-how-big-a-problem-are-they.

satellite internet on the go: Wendland, Mike. "Everything You Need to Know
about Starlink Internet for RVers." *Camping World,* July 14, 2022. https://
blog.campingworld.com/the-rv-life/fulltiming/everything-you-need-to
-know-about-starlink-internet-for-rvers/.

Russia invading Ukraine: Sheetz, Michael. "About 150,000 People in Ukraine Are
Using SpaceX's Starlink Internet Service Daily, Government Official Says."
CNBC, May 2, 2022. https://www.cnbc.com/2022/05/02/ukraine-official
-150000-using-spacexs-starlink-daily.html.

in thirty countries: Keane, Sean. "Musk's Starlink Internet Is Now Available in
32 Countries." *CNET,* May 13, 2022. https://www.cnet.com/home/internet
/starlink-internet-is-available-in-32-countries/.

2,500 satellites: Clark, Stephen. "SpaceX Passes 2,500 Satellites Launched for Starlink
Internet Network." *Spaceflight Now,* May 13, 2022. https://spaceflightnow
.com/2022/05/13/spacex-passes-2500-satellites-launched-for-companys-starlink
-network/.

"Formerly quiet residential streets": Paul, Pamela. *100 Things We've Lost to the
Internet.* United States: Crown, 2021, 22.

a team of scientists from the Czech Republic: Hejtmánek, Lukáš, Ivana Oravcová,

Jiří Motýl, Jiří Horáček, and Iveta Fajnerová. "Spatial Knowledge Impairment after GPS Guided Navigation." *International Journal of Human-Computer Studies*, April 22, 2018. https://www.sciencedirect.com/science/article/abs /pii/S107158191830171X.

"It came from wanting": Lia Li, interview with the author, April 20, 2022.

According to the National Transportation Safety Board: Phillips, Don. "NTSB Says Disorientation Likely Caused JFK Jr. Crash." *Washington Post*, July 7, 2000. https://www.washingtonpost.com/archive/politics/2000/07/07/ntsb-says -disorientation-likely-caused-jfk-jr-crash/08cd60a8-74ae-46e1-a2e8-960ab2e 71116/.

"I have a compass": Calaprice, Alice. *The Ultimate Quotable Einstein*. Princeton, NJ: Princeton University Press, 2013, 450.

EINSTEIN AND E.T.

1.8 million people: "U.S. Census Bureau Quickfacts: West Virginia." Accessed August 14, 2022. https://www.census.gov/quickfacts/WV.

world's largest teapot: "World's Largest Teapot, Chester, West Virginia." *RoadsideAmerica.com*. Accessed August 14, 2022. https://www.roadsideamerica .com/story/11259.

life-size replica of Noah's Ark: "Noah's Ark Being Rebuilt." *RoadsideAmerica.com*. Accessed August 14, 2022. https://www.roadsideamerica.com/story/10061.

America's smallest post office: "Silver Lake, WV—Smallest Post Office." *RoadsideAmerica.com*. Accessed August 14, 2022. https://www.roadsideamerica .com/tip/30236.

population seventy-four: "Census Profile: Green Bank, WV." *Census Reporter*. Accessed August 14, 2022. https://censusreporter.org/profiles/16000US5433124 -green-bank-wv/.

town's most famous export was Bruce Bosley: Hamill, Geoff. "Bruce Bosley—A Hometown Hero Worth Remembering." *Pocahontas Times*, June 4, 2014. https:// pocahontastimes.com/bruce-bosley-a-hometown-hero-worth-remembering/.

13,000-square-mile area: Levin, Dan, and Annie Flanagan. "No Cell Signal, No Wi-Fi, No Problem. Growing up inside America's 'Quiet Zone.'" *New York Times*, March 6, 2020. https://www.nytimes.com/2020/03/06/us/green-bank-west -virginia-quiet-zone.html.

They suffer from electromagnetic hypersensitivity: Stromberg, Joseph. "The Tiny Town Where the 'Electrosensitive' Can Escape the Modern World." *Slate*,

April 12, 2013. https://slate.com/technology/2013/04/green-bank-w-v-where
-the-electrosensitive-can-escape-the-modern-world.html.

"a washbasin for Godzilla": Kurczy, Stephen. *The Quiet Zone: Unraveling the Mystery
of a Town Suspended in Silence*. New York: Dey Street Books, 2021, 9.

The locals refer to the GBT: "The Great Big Thing." National Radio Astronomy
Observatory, November 19, 2020. https://public.nrao.edu/gallery/the-great
-big-thing/.

the longest-serving member: "A Tribute to Senator Robert C.
Byrd." National Archives and Records Administration. Accessed
August 14, 2022. https://www.archives.gov/legislative
/features/byrd.

opened for scientific exploration: "History of Green Bank and the NRAO."
National Radio Astronomy Observatory. Accessed August 14, 2022.
https://science.nrao.edu/facilities/gbt/green-bank-local-area-information
/history-of-green-bank-and-the-nrao.

"It is a pity": "Einstein and Mars." NASA. Accessed August 14, 2022. https://www
.jpl.nasa.gov/images/pia20811-einstein-and-mars.

He had just signed his divorce papers: Popova, Maria. "Einstein's Divorce
Agreement and the Nuanced Messiness of the Human Heart." *The Marginalian*,
July 16, 2016. https://www.themarginalian.org/2015/06/12/einstein-divorce/.

traveled to the island of Principe: Beléndez, Augusto. "The Eclipse to Confirm
the General Theory of Relativity." *OpenMind*, August 30, 2018. https://
www.bbvaopenmind.com/en/science/physics/the-eclipse-to-confirm
-the-general-theory-of-relativity/.

six minutes: Overbye, Dennis. "The Eclipse That Revealed the Universe." *New
York Times*, July 31, 2017. https://www.nytimes.com/2017/07/31/science
/eclipse-einstein-general-relativity.html.

sent their golf reporter: Gates, S. James, and Cathie Pelletier. *Proving Einstein Right:
The Daring Expeditions That Changed How We Look at the Universe*. New York:
PublicAffairs, 2019, 252.

declared in an all-caps headline: "Einstein Theory Triumphs." *New York Times*.
Accessed August 14, 2022. https://timesmachine.nytimes.com/timesmachine
/1919/11/10/118180487.html?pageNumber=17.

"barely breathe": Parkin, Simon. "Who Owns Einstein? The Battle for the World's
Most Famous Face." *The Guardian*. Guardian News and Media, May 17, 2022.
https://www.theguardian.com/media/2022/may/17/who-owns-einstein
-the-battle-for-the-worlds-most-famous-face.

"It's hard to think": Daniel Kennefick, phone interview with the author, April 29, 2019.

"In an infinite universe": Merali, Zeeya. "Search for Extraterrestrial Intelligence Gets a $100-Million Boost." *Nature*, July 20, 2015. https://www.nature.com /articles/nature.2015.18016.

Yuri Milner: "Biggest Science Prize Takes Web Tycoon from Social Networks to String Theory." *The Guardian*, July 31, 2012. https://www.theguardian.com /science/2012/jul/31/prize-science-yuri-milner-awards.

with a personal gift: Rundle, Michael. "$100m Breakthrough Listen Is 'Largest Ever' Search for Alien Civilisations." *Wired UK*, July 20, 2015. https://www .wired.co.uk/article/breakthrough-listen-project.

Harvard's Avi Loeb: Lochhead, Christopher. "Harvard's Top Astronomer Says Aliens Tried to Contact Us." December 21, 2021. https://lochhead.com /avi-loeb/.

a parrot at the Knoxville Zoo: "Einstein, Zoo Knoxville's African Grey Parrot, Shows off Its Intelligence in New Video." *WBIR*, July 29, 2020. https:// www.wbir.com/article/entertainment/places/zoo-knoxville/einstein-zoo -knoxvilles-african-grey-parrot-shows-off-its-big-birdbrain-in-new-video/51 –685a8b5a-861c-485f-a2a8–47230a863d53.

the *New York Times*: Overbye, Dennis. "Did an Alien Life-Form Do a Drive-by of Our Solar System in 2017?" *New York Times*, January 26, 2021. https://www .nytimes.com/2021/01/26/books/review/extraterrestrial-avi-loeb.html.

the *Wall Street Journal*: "Is That an Alien Probe? A Harvard Astronomer Thinks It Might Be." *Wall Street Journal*, February 9, 2019. https://www.wsj.com/articles /is-that-an-alien-probe-a-harvard-astronomer-thinks-it-might-be-11549638889.

and CBS News: "Author Avi Loeb on Evidence of Close Extraterrestrial Encounters, New Book." CBS News, March 1, 2021. https://www.cbsnews.com/video /author-avi-loeb-on-evidence-of-close-extraterrestrial-encounters-new-book/#x.

"I'm doing astronomy": Avi Loeb, interview with the author, February 19, 2018.

publish a peer-reviewed paper: Loeb, Avi, and Avery Broderick. "Imaging the Black Hole Silhouette of M87." *The Astrophysical Journal*, May 11, 2009. https://lweb .cfa.harvard.edu/~loeb/BL_2009.pdf.

gained international notoriety: Information@eso.org. "Astronomers Reveal First Image of the Black Hole at the Heart of Our Galaxy." European Southern Observatory. Accessed August 14, 2022. https://www.eso.org/public/news /eso2208-eht-mw/.

"That's not an impressive feat": *"Black Holes: The Edge of All We Know."* IMDb, 2020. https://www.imdb.com/title/tt11863046/.

bright ring of light: "'The Dawn of a New Era in Astronomy.'" Center for Astrophysics. Accessed August 14, 2022. https://pweb.cfa.harvard.edu/news/dawn-new-era-astronomy.

"a finely whiskered vortex": Overbye, Dennis. "The Most Intimate Portrait Yet of a Black Hole." *New York Times*, March 24, 2021. https://www.nytimes.com/2021/03/24/science/astronomy-messier-87-black-hole.html.

"the discovery of my lifetime": *"Black Holes: The Edge of All We Know."*

surprisingly decent TripAdvisor rating: "Area 51 Alien Travel Center, Amargosa Valley." Tripadvisor. Accessed August 14, 2022. https://www.tripadvisor.com/Restaurant_Review-g29719-d8026088-Reviews-Area_51_Alien_Travel_Center-Amargosa_Valley_Nevada.html.

"the FBI showed up": Gooden, Lezla. "Man Who Created 'Area 51' Event Says He Was Contacted by FBI." WFTS, August 8, 2019. https://www.abcactionnews.com/news/national/creator-of-viral-storm-area-51-facebook-event-says-he-was-contacted-by-fbi.

"I didn't feel comfortable": Knowles, Hannah. "'Storm Area 51' Creator Pulls out of His Own Event, Calling It Fyre Festival 2.0." *Washington Post*, October 23, 2019. https://www.washingtonpost.com/technology/2019/09/11/storm-area-creator-pulls-out-his-own-event-calling-it-fyre-festival/.

"It's a super heavy element": "Bob Lazar Says the FBI Raided Him to Seize Area 51's Alien Fuel. The Truth Is Weirder." *Vice*, November 13, 2019. https://www.vice.com/en/article/evjwkw/bob-lazar-says-the-fbi-raided-him-to-seize-area-51s-alien-fuel-the-truth-is-weirder.

NASA helped director Ridley Scott: McCarthy, Erin. "How NASA and Ridley Scott Collaborated to Make The Martian." *Mental Floss*, October 2, 2015. https://www.mentalfloss.com/article/69351/how-nasa-and-ridley-scott-collaborated-make-martian.

a fan of author Aaron Bernstein: "This Month in Physics History." *American Physical Society*, https://aps.org/publications/apsnews/200503/history.cfm.

philosophy by Immanuel Kant: Howard, Don A. "Albert Einstein as a Philosopher of Science." *Physics Today*, December 1, 2005. https://physicstoday.scitation.org/doi/10.1063/1.2169442.

Einstein and Wells would eventually meet: "When Einstein Met H.G. Wells." *Big Think*, November 5, 2021. https://bigthink.com/starts-with-a-bang/when-einstein-met-h-g-wells/.

"Hubble gave Einstein": Abraham, *Possessing Genius: The Bizarre Odyssey of Einstein's Brain*, 164.

"every reason to believe": Calaprice, Alice. *The Ultimate Quotable Einstein.* Princeton, NJ: Princeton University Press, 2013, 429.

a conclave on the topic: Wenz, John. "The Order of the Dolphin: Seti's Secret Origin Story." Astronomy.com, October 10, 2018. https://astronomy.com/news/2018/10/the-order-of-the-dolphin-setis-secret-origin-story.

they beamed a three-minute audio message: "Arecibo Message." SETI Institute. Accessed August 15, 2022. https://www.seti.org/seti-institute/project/details/arecibo-message.

Songs of humpback whales: "Voyager—Music on the Golden Record." NASA. Accessed August 15, 2022. https://voyager.jpl.nasa.gov/golden-record/whats-on-the-record/music/.

been burned onto a CD: "Various—Voyager Golden Record—Ozma Records." Amazon. Accessed August 15, 2022. https://www.amazon.com/Various-Voyager-Golden-Records-Ozma-001/dp/B0789V564H.

falling from above the tree line: "Podcast: Arecibo Observatory." *Atlas Obscura,* November 11, 2021. https://www.atlasobscura.com/articles/podcast-arecibo-observatory.

"My philosophy is more like": Worrall, Simon. "Buzz Aldrin Hates Being Called the Second Man on the Moon." *National Geographic,* April 18, 2016. https://www.nationalgeographic.com/science/article/160418-buzz-aldrin-ufo-apollo-crew-moon-mars-space-ngbooktalk.

"Our death is not an end": Calaprice, Alice. *The Ultimate Quotable Einstein.* Princeton, NJ: Princeton University Press, 2013, 91.

Einstein first described in a 1917 paper: "This Month in Physics History." American Physical Society. Accessed August 15, 2022. https://www.aps.org/publications/apsnews/200508/history.cfm.

launched the world's smallest spacecraft: "Breakthrough Starshot Successfully Launch World's Smallest Spacecraft." *The Guardian,* July 28, 2017. https://www.theguardian.com/science/2017/jul/28/breakthrough-starshot-successfully-launch-worlds-smallest-spacecraft.

"a natural presumption": Calaprice, Alice. *The Ultimate Quotable Einstein.* Princeton, NJ: Princeton University Press, 2013, 79.

got an unexpected jolt: Gazette-Mail, Rick Steelhammer. "West Virginia's Green Bank Scope Tracks Object for Signs of Alien Design." *WV News,* December 13, 2017. https://www.wvnews.com/news/wvnews/west-virginia-s-green-bank-scope-tracks-object-for-signs-of-alien-design/article_ee928875-552e-50b2-8cf1-2b51aa3b5b25.html.

The scientists dubbed it "Oumuamua": Chotiner, Isaac. "Have Aliens Found Us? A Harvard Astronomer on the Mysterious Interstellar Object 'Oumuamua." *New Yorker*, January 16, 2019. https://www.newyorker.com/news/q-and-a /have-aliens-found-us-a-harvard-astronomer-on-the-mysterious-interstellar -object-oumuamua.

he had spotted a second: Tress, Luke. "Israeli Astronomer and Partner Identify First Interstellar Meteor to Hit Earth." *Times of Israel*, April 17, 2022. https:// www.timesofisrael.com/israeli-astronomer-and-partner-identify-first -interstellar-meteor-to-hit-earth/.

donors offered Loeb $1.7 million: Mann, Adam. "Avi Loeb's Galileo Project Will Search for Evidence of Alien Visitation." *Scientific American*, July 27, 2021. https://www.scientificamerican.com/article/avi-loebs-galileo-project -will-search-for-evidence-of-alien-visitation/.

which skyrocketed up: "Combined Print & E-Book Nonfiction—Best Sellers— Books—Feb. 14, 2021." *New York Times*. Accessed August 15, 2022. https:// www.nytimes.com/books/best-sellers/2021/02/14/combined-print-and-e -book-nonfiction/.

not without controversy: McMahon, Sean. "Do Extraordinary Claims Require Extraordinary Evidence? The Proper Role of Sagan's Dictum in Astrobiology." Oxford University Press, May 25, 2020. https://academic.oup.com/book/36896 /chapter-abstract/322138781.

A 1931 book: Lopate, Phillip. "One Hundred Authors against Einstein." Amazon. Accessed August 15, 2022. https://www.amazon.com/One-Hundred-Authors -Against-Einstein/dp/B09PHH7KC8.

two petabytes of data: "Breakthrough Listen Releases 2 Petabytes of Data from SETI Survey of Milky Way." *Phys.org*, February 15, 2020. https://phys.org /news/2020-02-breakthrough-petabytes-seti-survey-milky.html.

"so far since 1960": Kramer, Miriam. "The Project Dedicated to Finding Extraterrestrial Life Has Only Scratched the Surface." *Axios*, June 25, 2019. https://www.axios.com/2019/06/25/life-space-breakthrough-listen.

dropped the largest data dump of its kind: Cowing, Keith. "Breakthrough Listen Publishes Most Comprehensive and Sensitive Search for Radio Technosignatures Ever Performed." *Astrobiology*, June 18, 2019. https:// astrobiology.com/2019/06/breakthrough-listen-publishes-most -comprehensive-and-sensitive-search-for-radio-technosignatures-eve.html.

"looking to conquer and colonize": Payton, Matt. "Aliens Could Come to Earth to 'Conquer and Colonise'—Says Stephen Hawkings." *Metro*, December 11, 2019.

https://metro.co.uk/2015/10/02/aliens-could-come-to-earth-to-conquer -and-colonise-says-stephen-hawkings-5417652/.

sniffed out methane: Shekhtman, Svetlana. "NASA Scientists Closer to Explaining Mars Methane Mystery." NASA, May 26, 2021. https://www.nasa.gov/feature /goddard/2021/first-you-see-it-then-you-don-t-scientists-closer-to-explaining -mars-methane-mystery/.

called "gravitational microlensing": Rothenberg Gritz, Jennie. "The Wonder of Avi Loeb." *Smithsonian*, October 1, 2021. https://www.smithsonianmag.com /science-nature/wonder-avi-loeb-180978579/.

Carl Sagan famously said: Koupelis, Theo. *In Quest of the Universe*. United States: Jones & Bartlett Learning, 2010, 551.

WAR AND PEACE...AND EINSTEIN

"I say the same thing": Mandy Patinkin, interview with the author, January 7, 2022.

"Take a very good look at it": Isaacson, Walter. *Einstein: His Life and Universe*. United Kingdom: Simon & Schuster UK, 2008, 401.

The Hitler Youth: "Paradise Lost." *New Scientist*, January 21, 1995. https://www .newscientist.com/article/mg14519612-100-paradise-lost/.

a $5,000 bounty: Ciaccia, Chris. "Einstein Letters Revealing Escape from Nazis up for Auction." *New York Post*, June 27, 2018. https://nypost.com/2018/06/27 /einstein-letters-revealing-escape-from-nazis-up-for-auction/.

"I didn't know I was worth so much": Jerome, Fred. *The Einstein File: J. Edgar Hoover's Secret War against the World's Most Famous Scientist*. United States: St. Martin's Press, 2003, 24.

Melania Trump: Jordan, Miriam. "Did Melania Trump Merit an 'Einstein Visa'? Probably, Immigration Lawyers Say." *New York Times*, March 4, 2018. https:// www.nytimes.com/2018/03/04/us/melania-trump-einstein-visa.html.

performed a violin concert: Katz, Mandy. "Was Einstein a Jewish Saint?" *Moment*, June 5, 2017. https://momentmag.com/was-einstein-a-jewish-saint/.

Jews in Alaska and Mexico: Parkin, Simon. "Who Owns Einstein? The Battle for the World's Most Famous Face." *The Guardian*, May 17, 2022. https:// www.theguardian.com/media/2022/may/17/who-owns-einstein-the-battle -for-the-worlds-most-famous-face.

"I am privileged by fate": Calaprice, Alice. *The Ultimate Quotable Einstein*. Princeton, NJ: Princeton University Press, 2013, 72.

In the summer of 1933: "Albert Einstein and the Birth of the International Rescue

Committee." The IRC. Accessed August 20, 2022. https://www.rescue.org
/article/albert-einstein-and-birth-international-rescue-committee.

by 85 percent: Watson, Julie. "Broken by Trump, US Refugee Program Aims to Return
Stronger." *Associated Press*, January 27, 2021. https://apnews.com/article/joe
-biden-politics-immigration-coronavirus-pandemic-0a649290b8a6628900598
d4324c3d72b.

"Historically, refugees were in camps": George Biddle, interview with the author,
December 13, 2021.

opened their source code: "IRC Open-Sources Digital Tracking System for
International Aid," August 12, 2015. https://philanthropynewsdigest.org/news
/irc-open-sources-digital-tracking-system-for-international-aid.

Einstein opened the doors: Lombardo, Cristiana. "Marian Anderson and Albert
Einstein's Unexpected Friendship." *PBS*, June 7, 2022. https://www.pbs.org
/wnet/americanmasters/marian-anderson-and-albert-einsteins-unexpected
-friendship/20766/.

"My attitude is not derived": Calaprice, Alice. *The Ultimate Quotable Einstein.*
Princeton, NJ: Princeton University Press, 2013, 247.

commencement address at Lincoln University: Phillips, Kristine. "Albert
Einstein Decried Racism in America. His Diaries Reveal a Xenophobic,
Misogynistic Side." *Washington Post*, October 28, 2021. https://www
.washingtonpost.com/news/retropolis/wp/2018/06/13/albert-einstein
-decried-racism-in-america-his-diaries-reveal-a-xenophobic-misogynistic-side/.

"The more I feel an American": Taghavi, Aram Rasa. "Albert Einstein in His Famous
1946 Essay on Race: 'The Negro Question.'" *Medium*, October 9, 2018. https://
medium.com/@ARTaghavi/albert-einstein-in-his-famous-1946-essay-on-race
-the-negro-question-555f9a846b4e#:~:text=The%20more%20I%20feel
%20an,with%20Negroes%20in%20this%20country.

"Because he was so famous": Michael Gordin, interview with the author, March
30, 2022.

Phillippe Halsman: Waxman, Olivia B. "The Story of Albert Einstein's 'Magnificent
Birthday Gift.'" *Time*, March 13, 2020. https://time.com/5795646/albert
-einstein-birthday-pi-day/.

more than 10 million Ukrainians fled: "Border Crossings from Ukraine since War
Began Passes 10 MLN Mark UN Agency." *Reuters*, August 2, 2022. https://
www.reuters.com/world/europe/border-crossings-ukraine-since-war-began
-passes-10-mln-mark-un-agency-2022-08-02/.

"I had a feeling": Anna Pysana, interview with the author, May 2, 2022.

was the first city seized: "Ukraine Crisis: Casualties in Sloviansk Gun Battles." *BBC News*, April 13, 2014. https://www.bbc.com/news/world-europe-27008026.

The majority of them emigrated to Israel: "How One Million Russian Immigrants Changed the Face of Israel." *Haaretz*. Accessed August 21, 2022. https://www.haaretz.com/st/c/prod/eng/25yrs_russ_img/.

one of six in Ukraine: Michael Geller, interview with the author, April 27, 2022.

he attended a meeting: "Einstein Appeals to U.S. Jews to Continue Relief Work of Joint Distribution Committee." *Jewish Telegraphic Agency*, December 16, 1930. https://www.jta.org/archive/einstein-appeals-to-u-s-jews-to-continue-relief-work-of-joint-distribution-committee.

"a new lease on life": Waxman, Olivia B. "The Story of Albert Einstein's 'Magnificent Birthday Gift.'" *Time*, March 13, 2020. https://time.com/5795646/albert-einstein-birthday-pi-day/.

A 1946 cover of *Time* magazine: "Time Magazine Cover: Albert Einstein—July 1, 1946." *Time*. Accessed August 21, 2022. http://content.time.com/time/covers/0,16641,19460701,00.html.

"My participation in the production": Calaprice, Alice. *The Ultimate Quotable Einstein*. Princeton, NJ: Princeton University Press, 2013, 258.

"There was never": Ibid., 284.

"The way I like to put it": Wallerstein, Alex. "A Bomb without Einstein?" Restricted Data: The Nuclear Secrecy Blog, June 27, 2014. http://blog.nuclearsecrecy.com/2014/06/27/bomb-without-einstein/.

"The use of the bomb": Feld, Bernard T. "Einstein and the Politics of Nuclear Weapons" In *Albert Einstein, Historical and Cultural Perspectives: The Centennial Symposium in Jerusalem*, edited by Gerald Holton and Yehuda Elkana, 369–394. Princeton: Princeton University Press, 2014. https://doi.org/10.1515/9781400855438.369.

"I am not only a pacifist": Calaprice, *The Ultimate Quotable Einstein*, 252.

"members of this or that nation": "Russell-Einstein Manifesto." Atomic Heritage Foundation. Accessed August 21, 2022. https://www.atomicheritage.org/key-documents/russell-einstein-manifesto.

EINSTEIN'S MIRACLE YEAR

a connection from the father of a friend: Violaris, Maria. "Einstein at the Patent Office." *The Oxford Scientist*, April 26, 2020. https://oxsci.org/einstein-at-the-patent-office/.

"His visual imagination": Isaacson, Walter. *Einstein: His Life and Universe*. United States: Simon & Schuster, 2017, 93.

"He was an ambassador": Abraham, *Possessing Genius: The Bizarre Odyssey of Einstein's Brain*, 318.

his second-floor apartment: "Einstein House in Bern—Apartment Einstein and His Wife Mileva Lived in 1905." Switzerland by Rail. Accessed August 16, 2022. http://www.switzerlandbyrail.com/destinations/bern/einstein_house.htm.

slept in the other room: Overbye, Dennis. *Einstein in Love: A Scientific Romance*. United States: Penguin Publishing Group, 2001, 110.

sells T-shirts and hoodies: Shop Einstein. Accessed August 16, 2022. https://shopeinstein.com/.

called him a "lazy dog": Gribbin, John. "Pay Attention, Albert Einstein!" *New Scientist*, January 2, 1993. https://www.newscientist.com/article/mg13718543-900-pay-attention-albert-einstein/.

When two black holes merge: "LIGO's Dual Detectors." Caltech. Accessed August 16, 2022. https://www.ligo.caltech.edu/page/ligo-detectors.

Barry Barish, Kip Thorne, and Ranier Weiss: "LIGO Pioneers Named 2017 Nobel Prize in Physics Laureates." NSF. Accessed August 16, 2022. https://www.nsf.gov/news/news_summ.jsp?cntn_id=243280.

largest projects ever funded: "Gravitational Waves Detected 100 Years after Einstein's Prediction." Caltech. Accessed August 16, 2022. https://www.ligo.caltech.edu/news/ligo20160211.

"That simple chirp": Overbye, Dennis. "Gravitational Waves Detected, Confirming Einstein's Theory." *New York Times*, February 11, 2016. https://www.nytimes.com/2016/02/12/science/ligo-gravitational-waves-black-holes-einstein.html.

"the most powerful thing": "The Man Who Helped Prove Einstein Correct Weighs in on America's Startling Science Gap." *Los Angeles Times*, November 15, 2017. https://www.latimes.com/opinion/op-ed/la-ol-patt-morrison-kip-thorne-nobel-science-20171115-htmlstory.html.

second pair of merging black holes: "LIGO Detects Another Black Hole Crash." *Science*. Accessed August 16, 2022. https://www.science.org/content/article/ligo-detects-another-black-hole-crash.

The scientists published a paper: "Observation of Gravitational Waves from a Binary Black Hole Merger." Accessed August 16, 2022. https://authors.library.caltech.edu/64405/1/PhysRevLett.116.061102.pdf.

they were awarded the Nobel Prize in Physics: "2017 Nobel Prize in Physics

Awarded to LIGO Founders." Caltech. Accessed August 16, 2022. https://www
.ligo.caltech.edu/page/press-release-2017-nobel-prize.

laid the groundwork for so many technologies: Howell, Elizabeth. "Photoelectric
Effect: Explanation & Applications." *LiveScience*, April 25, 2017. https://www
.livescience.com/58816-photoelectric-effect.html.

"little ball-like objects": Sean McWilliams, interview with the author, June 2, 2022.

Japanese scientists announced: Luntz, Stephen. "Quantum Phenomenon Used
to Kill Cancers." *IFLScience*, July 14, 2021. https://www.iflscience.com
/quantum-phenomenon-used-to-kill-cancers-for-real-for-once-60349.

"Every photoelectric cell can be considered": Holton, Gerald, and Yehuda Elkana.
*Albert Einstein, Historical and Cultural Perspectives: The Centennial Symposium in
Jerusalem*. Princeton, NJ: Princeton University Press, 2016, vii.

"And that gum": Mintz, Corey. "Why Grocery Shopping Is on Its Way Out." *The Walrus*,
April 10, 2022. https://thewalrus.ca/why-grocery-shopping-is-on-its-way-out/.

Goldman hired male actors: Dunne, Carey. "Weird History: Inventor Hired Models
to Make Shopping Carts Seem Cool." *Fast Company*, August 12, 2014. https://
www.fastcompany.com/3034248/weird-history-inventor-hired-models
-to-make-shopping-carts-seem-cool.

Weather prediction: Nordam, Tor. "Not so Random after All? How Random
Numbers Help Us to Predict the Physical World." SINTEFblog, March 18, 2022.
https://blog.sintef.com/sintefocean/not-so-random-after-all-how-random
-numbers-help-us-to-predict-the-physical-world/.

can also be applied to the stock market: Ermogenous, Angeliki. "Brownian
Motion and Its Applications in the Stock Market." Accessed August 18, 2022.
https://ecommons.udayton.edu/cgi/viewcontent.cgi?article=1010&context
=mth_epumd.

the suspension remains: "Technological Strategies to Estimate and Control
Diffusive Passage Times through the Mucus Barrier in Mucosal Drug Delivery."
U.S. National Library of Medicine, January 15, 2018. https://www.ncbi.nlm
.nih.gov/pmc/articles/PMC5809312/.

"Einstein took every opportunity to disavow it": Holton, Gerald. *Einstein, History,
and Other Passions: The Rebellion Against Science at the End of the Twentieth
Century*. Cambridge, MA: Harvard University Press, 2000, 129.

"it is a puzzle in itself": Einstein, Albert. *Einstein's 1912 Manuscript on the Special
Theory of Relativity*. New York: George Braziller, 2003, 11.

"The contemplation is amusing": Calaprice, Alice. *The Ultimate Quotable Einstein*.
Princeton, NJ: Princeton University Press, 2013, 355.

take a flashlight: Greene, Brian. "Your Daily Equation #1: E = MC2." YouTube. World Science Festival, March 26, 2020. https://youtu.be/G_NlwfA8x0o.

It sold for $6.5 million: Hampson, Rick. "Einstein Manuscript Commands Relatively High Price." Associated Press, December 2, 1987. https://apnews.com/article/5f6c321686612d66fbea085d03732a09.

When a plague ravaged Europe: Cohen, Ben. "How the Plague Ravaged Shakespeare's World and Inspired His Work." *Slate Magazine*. Slate, March 10, 2020. https://slate.com/culture/2020/03/shakespeare-plague-influence-hot-hand-ben-cohen.html.

Between 1914 and 1918: Stanley, Matthew. *Einstein's War: How Relativity Triumphed Amid the Vicious Nationalism of World War I.* United Kingdom: Dutton, 2019, 3.

"There is something to this idea": Ben Cohen, interview with the author, June 11, 2020.

cement mixing to deodorant: Gribbin, John. "The Everyday World of Einstein." *New Scientist*, December 25, 1993. https://www.newscientist.com/article/mg14019054-400-the-everyday-world-of-einstein-what-did-albert-want-with-a-cup-of-sweet-coffee-a-cement-mixer-and-a-dirty-cloud/.

why the sky is blue: "Why Is the Sky Blue?" Culture.pl. Accessed August 19, 2022. https://culture.pl/en/article/why-is-the-sky-blue-the-polish-scientist-who-found-the-answer.

BRAND EINSTEIN

holds the Guinness World Record: Kingsbury, Alex. "The Autograph Alternative: Human Hair." CBS News, October 30, 2007. https://www.cbsnews.com/news/the-autograph-alternative-human-hair/.

a small clump of Elvis's hair: CBSnews.com. "Auctioned Elvis Hair Fetches $15,000." CBS News, October 19, 2009. https://www.cbsnews.com/news/auctioned-elvis-hair-fetches-15000/.

Michael Jackson's hair: NPR. "Collector Guards His Golden Locks." NPR, October 18, 2009. https://www.npr.org/templates/story/story.php?storyId=113916955.

who had given a buzz to Neil Armstrong: NBC Universal. "Astronaut's Hair Sparks Legal Hubbub." NBC News, June 2005. https://www.nbcnews.com/id/wbna8062442.

"It's a very Victorian tradition": John Reznikoff, interview with the author, June 21, 2021.

the world's most expensive bottle of wine: Wallace, Benjamin. *The Billionaire's Vinegar: The Mystery of the World's Most Expensive Bottle of Wine.* United States: Three Rivers Press, 2009.

The Forbes family bought: Brown, Abram. "Jewels, Eggs and Empires: The Story of Forbes and Faberge." *Forbes*, May 9, 2018. https://www.forbes.com/sites /abrambrown/2017/09/19/forbes-faberge/?sh=17a786ca4ccf.

so much Chinese art: Keefe, Patrick Radden. *Empire of Pain: The Secret History of the Sackler Dynasty.* United States: Knopf Doubleday Publishing Group, 2021.

"There's enough George Washington hair": Bill Panagalopolus, interview with the author, April 11, 2011.

"It would almost be remiss": O'Neal, Sean. "Daily Buzzkills: Michael Jackson's Hair-Diamonds Are Forever." *The A.V. Club.* The A.V. Club, July 30, 2009. https://www.avclub.com/daily-buzzkills-michael-jacksons-hair-diamonds-are -for-1798217233.

The 1987 auction of a handwritten paper: Reif, Rita. "Einstein Paper Sets Auction Record."*New York Times*, December 3, 1987. https://www.nytimes.com/1987 /12/03/books/einstein-paper-sets-auction-record.html.

X-ray of his skull: Julienslive.com. "ALBERT EINSTEIN BRAIN X-RAYS," 2022. https://www.julienslive.com/lot-details/index/catalog/11/lot/2364.

a batch of Einstein's letters: Cleveland, Lauriel. "Einstein Letters Fetch More than $420,000 at Auction." CNN, June 15, 2015. https://www .cnn.com/2015/06/15/us/einstein-letters-auction.

An autographed copy of the classic photo: Gannon, Megan. "Iconic Photo of Einstein Sticking out His Tongue Sells for $125,000." Live Science, July 29, 2017. https://www.livescience.com/59978-einstein-tongue -photo-sells-at-auction.html.

A violin once owned: Hanlon, Mike. "Einstein's Violin Fetches $516,500 at New York Auction." New Atlas, March 11, 2018. https://newatlas.com/einstein-violin -auction/53760/.

auction of eight letters: Timesofisrael.com. "Einstein Letters on God, Israel, Physics Fetch $210,000," June 21, 2017. https://www.timesofisrael.com/einstein-letters -on-god-israel-and-physics-fetch-210000/.

mentalist Uri Geller: "Einstein Letters on God, Israel and Physics Fetch $210,000." *Phys.org*, June 20, 2017. https://phys.org/news/2017-06-einstein-letters-god -israel-physics.html.

That wax mold: University Archives. "Albert Einstein Signs a Wax Mold Casted Off His

Face 'from Life.'" 2022. https:/auction.universityarchives.com/auction-lot/albert
-einstein-signs-a-wax-mold-head-casted-off_80540F6AF6.

Einstein's doctoral dissertation: Universityarchives.com. "Lot—Albert Einstein
Signed Copy of 2 Important Works Incl. His Doctoral Thesis, 'a New
Determination of Molecular Dimensions'—Extraordinarily Rare!" 2020.
https://auction.universityarchives.com/auction-lot/albert-einstein-signed
-copy-of-2-important-works_6774187A57.

The genius's iconic pipe: Christies.com. "[EINSTEIN, Albert (1879–1955)]."
Christie's, 2020. https://www.christies.com/en/lot/lot-6089296.

Einstein's prediction came true: Bilesky, Dan. "Albert Einstein's 'Theory of
Happiness' Fetches $1.56 Million," *New York Times*, October 25, 2017. https://
www.nytimes.com/2017/10/25/world/middleeast/einstein-theory-of-
happiness.html.

a rare fifty-four-page manuscript: Cohen, Benyamin. "Einstein Manuscript Goes
for $11 Million at Auction, Highest Ever Paid for the Genius' Memorabilia,"
The Forward, November 23, 2021. https://forward.com/news/478616
/einstein-manuscript-goes-for-11-million-besso-christies-auction-relativity/.

in an auction for charity on eBay: Derschowitz, Jessica. "Justin Bieber's Hair
Sells for $40,668 on EBay." CBS News, March 2, 2011. https://www
.cbsnews.com/news/justin-biebers-hair-sells-for-40668-on-ebay/.

dubbed this phenomenon "celebrity contagion": Tierney, John. "Urge to Own
That Clapton Guitar Is Contagious, Scientists Find." *New York Times,* March 8,
2011. https://www.nytimes.com/2011/03/09/science/09guitar.html#:~:text
=%E2%80%9COur%20results%20suggest%20that%20physical,Newman
%20and%20Gil%20Diesendruck.

"He came at the moment": Kean, Sam. "The Face of Science." The National
Endowment for the Humanities, May 2014. https://www.neh.gov/humanities
/2014/mayjune/feature/the-face-science.

"While lecturing on the cosmos at Caltech": Abraham, Carolyn. *Possessing Genius:
The Bizarre Odyssey of Einstein's Brain.* United Kingdom: Icon, 2004, 179.

"captured the zeitgeist": Klosterman, Chuck. *The Nineties: A Book.* United
States: Penguin Publishing Group, 2022, 31.

"Pardon me, so sorry": Maloney, Russell, and E. Libman. "Disguise." *New Yorker,*
January 7, 1939. https://www.newyorker.com/magazine/1939/01/14
/disguise-2.

"People used to eat people": *Albert Einstein: Creator and Rebel.* Kiribati: Plunkett
Lake Press, 2019.

thousands of Japanese fans: Wamsley, Laurel. "Einstein's Note on Happiness, Given to Bellboy in 1922, Fetches $1.6 Million." NPR, October 25, 2017. https://www.npr.org/sections/thetwo-way/2017/10/25/560004689/einsteins-note-on-happiness-given-to-bellboy-in-1922-fetches-1-6-million.

"into a kind of idol": Calaprice, Alice. *The Ultimate Quotable Einstein.* United States: Princeton University Press, 2019, 19.

"Churches displayed statues of him": Kean, Sam. "The Face of Science." The National Endowment for the Humanities, May 2014. https://www.neh.gov/humanities/2014/mayjune/feature/the-face-science.

"By unraveling mysteries of the universe": Abraham, *Possessing Genius: The Bizarre Odyssey of Einstein's Brain,* 159.

"The interest in Einstein does not fade": Hanoch Gutfreund, interview with the author, November 12, 2015.

"the Einstein of octopuses": Georgiou, Aristos. "Why the Octopus from the Oscar-Winning Documentary Is the 'Einstein of Octopuses.'" Newsweek, June 1, 2021. https://www.newsweek.com/my-octopus-teacher-oscar-best-documentary-feature-1595855.

"The fact that he is as famous as he is": Michael Gordin, interview with the author, March 30, 2022.

"Companies he had never heard of": Abraham, *Possessing Genius: The Bizarre Odyssey of Einstein's Brain,* 5.

The Walt Disney company paid: Parkin, Simon. "Who Owns Einstein? The Battle for the World's Most Famous Face." *The Guardian.* Guardian News and Media, May 17, 2022. https://www.theguardian.com/media/2022/may/17/who-owns-einstein-the-battle-for-the-worlds-most-famous-face.

Huggies wanted to make: Ibid.

had strong ties to the Third Reich: Rosenberg, Eli. "German Billionaire Family That Owns Einstein Bros.. Bagels Admits Nazi Past." *Washington Post,* March 25, 2019. https://www.washingtonpost.com/history/2019/03/25/german-billionaire-family-that-owns-einstein-bros-bagels-admits-nazi-past/.

"The best and worst leaders": Poniewozik, James. *Audience of One: Donald Trump, Television, and the Fracturing of America.* United States: Liveright, 2019.

"Einstein is a cool guy": Roni Grosz, interview with the author, April 2, 2018.

EINSTEIN AND TIME TRAVEL

"my greatest inspirations": Lloyd, Christopher. Facebook, March 26, 2018. https://

www.facebook.com/thechristopherlloyd/posts/pfbid02m5icMrrUUdFp
2pq6GkG2BYv53fi4C9HtFJnj2Xa6Wpd9RUMeftGeXaLpfiRLcrSjl.

Lloyd is actually distantly related: AJ Jacobs, interview with the author, June 7, 2021.

"I used to read about scientists like Einstein": "Christopher Lloyd Breaks down His Most Iconic Characters." YouTube. *GQ*, January 31, 2022. https://youtu.be /BJuuOh8UJzM.

"She started out with nothing": Ronald Mallett, interview with the author, July 29, 2021.

he had constructed a mathematical model: Mallett, Ronald. "Weak Gravitational Field of the Electromagnetic Radiation in a Ring Laser." *Physics Letters*, April 3, 2000. https://citeseerx.ist.psu.edu/viewdoc/download?doi =10.1.1.511.5618&rep=rep1&type=pdf.

But everything changed: "Issue 2291: Magazine Cover Date: 19 May 2001." *New Scientist*. Accessed August 22, 2022. https://www.newscientist.com/issue /2291/.

two seats on each flight for "Mr. Clock": James, C. Renee. "Like GPS? Thank Relativity." *Discover Magazine*, August 29, 2014. https://www.discovermagazine .com/the-sciences/like-gps-thank-relativity.

the most time in space: Drake, Nadia. "How 879 Days of Spaceflight Changed This Cosmonaut." *National Geographic*, 2018. https://www.nationalgeographic .com/science/article/yuris-night-cosmonaut-gennady-padalka-longest -human-space-science.

Some people blamed: Beall, Abigail. "CERN Deny That the Hadron Collider Caused Italy's Earthquakes after Bizarre Claims." *Daily Mail*, November 7, 2016. https://www.dailymail.co.uk/news/article-3913952/Were-Italy-s-earthquakes -caused-HADRON-COLLIDER-Bizarre-theory-emerges-experiment-fire -plasma-Geneva-250-miles-underground-Italy.html.

commented on Godol's work: Henderson, Bruce, and Dr. Ronald L. Mallet. *Time Traveler: A Scientist's Personal Mission to Make Time Travel a Reality*. United States: Basic Books, 2009, 49.

to sign on to a letter: "The Einstein Letter That Started It All; a Message to President Roosevelt 25 Years Ago Launched the Atom Bomb and the Atomic Age." *New York Times*, August 2, 1964. https://www.nytimes.com/1964/08/02 /archives/the-einstein-letter-that-started-it-all-a-message-to-president.html.

"When you invent the ship": DiscoverQuotes.com. Accessed August 22, 2022. https://discoverquotes.com/paul-virilio/quote3084641/.

"Greetings. I am a time traveler": Jensen, K. Thor. "The Oral History of John Titor, the Man Who Traveled Back in Time to Save the Internet." *Thrillist*, October 19, 2017. https://www.thrillist.com/entertainment/nation/john-titor-time -traveler-predictions-story.

sent back to 1975: Lange, Steve. "John Titor: Who Was the Time Traveler That Visited Rochester?" *Rochester Post Bulletin*, October 30, 2020. https:// www.postbulletin.com/rochester-magazine/john-titor-who-was-the-time -traveler-that-visited-rochester.

"stop COVID-19's patient zero": "Young Physicist 'Squares the Numbers' on Time Travel." *UQ News*, September 24, 2020. https://www.uq.edu.au/news /article/2020/09/young-physicist-squares-numbers%E2%80%99-time-travel.

"I think changing past events": Richmond, Alasdair. "Time Travel Testimony and the 'John Titor' Fiasco." Think. Cambridge University Press, September 28, 2010. https://www.cambridge.org/core/journals/think /article/abs/time-travel-testimony-and-the-john-titor-fiasco/B673DD 9C1E8A741E23A64937AE0AA426.

He hosted a cocktail reception: "Stephen Hawking—Time Traveller's Party." YouTube, May 16, 2011. https://youtu.be/elah3i_WiFI.

"Time travelers are dicks": Wharton, David. "Five Possible Reasons Nobody Showed up for Stephen Hawking's Retroactive Time Traveler Party." *Giant Freakin Robot*, October 27, 2013. https://www.giantfreakinrobot.com/scifi /reasons-showed-stephen-hawkings-retroactive-time-traveler-party.html.

Two physics students at Michigan Technological University: Nemiroff, Robert J., and Teresa Wilson. "Searching the Internet for Evidence of Time Travelers." *arXiv.org*, December 26, 2013. https://arxiv.org/abs/1312.7128.

EINSTEIN LIFE HACKS

"When he couldn't find his sandals": Goodman, Elyssa. "Remember When Albert Einstein Invented Normcore?" *Yahoo!*, March 14, 2019. https://www.yahoo .com/now/remember-albert-einstein-invented-normcore-170000808.html.

"galactic clairvoyant": Abraham, *Possessing Genius: The Bizarre Odyssey of Einstein's Brain*, 10.

"urged him repeatedly to dress up": Ibid., 245.

"It is almost as if": Albert Einstein, *Historical and Cultural Perspectives: The Centennial Symposium in Jerusalem*. United States: Princeton University Press, 2014, 165.

"He was slovenly": Parkin, Simon. "Who Owns Einstein? The Battle for the World's Most

FamousFace." *TheGuardian*,May17,2022.https://www.theguardian.com/media /2022/may/17/who-owns-einstein-the-battle-for-the-worlds-most-famous-face.

"Uniform dressing has roots": Goodman, Elyssa. "Remember When Albert Einstein Invented Normcore?" *Yahoo!*, March 14, 2019. https://www.yahoo.com /now/remember-albert-einstein-invented-normcore-170000808.html.

"flying blind": Waldow, Herta, and Friedrich Herneck. *Einstein at Home*. United States: Prometheus Books, 2016.

"when one uses Einstein's handwriting": Harald Geisler, interview with the author, July 20, 2016.

mushrooms and two fried eggs: MacBride, Katie. "What Did Einstein Eat? Inside the Diet of the Famed Physicist." *Inverse*, March 14, 2022. https://www.inverse .com/mind-body/einstein-diet-brain-food-health.

an expandable shirt: *Patent Yogi*, October 21, 2015. https://patentyogi .com/event-of-the-day/einstein-patented-a-blouse/.

focused on something mundane: Colinio, Stacey. "The Science of Why You Have Great Ideas in the Shower." *National Geographic*, August 12, 2022. https://www.nationalgeographic.co.uk/history-and-civilisation/2022/08/the -science-of-why-you-have-great-ideas-in-the-shower.

"A cruise in the sea": Livni, Ephrat. "Albert Einstein's Best Ideas Came When He Was Aimless. Yours Can Too." *Quartz*, June 8, 2018. https://qz.com/1299282 /albert-einsteins-best-ideas-came-while-he-was-relaxing-aimlessly-yours-can -too/.

"According to his biographers": Devlin, Philip R. "Sailing the Connecticut Coast with Albert Einstein." *Patch*, March 21, 2013. https://patch.com/connecticut /mansfield/sailing-the-connecticut-coast-with-albert-einstein-5bbe11b8.

"He used to sail": Jim Lynch, interview with the author, June 27, 2016.

As Walter Isaacson noted: Isaacson, Walter. *Einstein: His Life and Universe*. United Kingdom: Simon & Schuster UK, 2008.

"aimlessly and often carelessly": Ibid., 435.

"the sweeping view": Calaprice, Alice. *The Ultimate Quotable Einstein*. Princeton, NJ: Princeton University Press, 2013, 12.

"Einstein's secret was that": Overbye, Dennis. "The Daily: Cosmic Questions." *New York Times*, August 19, 2022. https://www.nytimes.com/2022/08/19 /podcasts/the-daily/cosmos-space-black-holes.html.

"Answers can kill": Jacobs, A.J. "Alex Trebek: What I've Learned." *Esquire*, April 2003. https://classic.esquire.com/article/2003/4/1/alex-trebek.

the "puzzle mindset": Jacobs, A.J. *The Puzzler: One Man's Quest to Solve the Most*

Baffling Puzzles Ever, from Crosswords to Jigsaws to the Meaning of Life. United States: Crown, 2022.

four types of thinkers: Stillman, Jessica. "Adam Grant Explained 4 Modes of Thinking. There's 1 You Should Use Much More." *Inc.*, April 7, 2021. https://www.inc.com/jessica-stillman/there-are-4-modes-of-thinking-preacher-prosecutor-politician-scientist-you-should-use-1-much-more.html.

"favor humility over pride": Suttie, Jill. "Why Thinking Like a Scientist Is Good for You." *Greater Good*, March 24, 2021. https://greatergood.berkeley.edu/article/item/why_thinking_like_a_scientist_is_good_for_you.

an insurance salesman: Chung, Arthur. "Albert Einstein. His Struggles. His Failures." *Medium*, May 5, 2014. https://medium.com/@ArthurChung_/albert-einstein-his-struggles-his-failures-d7554f02b237.

handed out candy: Gewertz, Ken. "Albert Einstein, Civil Rights Activist." *Harvard Gazette*, April 12, 2007. https://news.harvard.edu/gazette/story/2007/04/albert-einstein-civil-rights-activist/.

a noiseless refrigerator: Kean, Sam. "Einstein's Little-Known Passion Project? A Refrigerator." *Wired*, July 23, 2017. https://www.wired.com/story/einsteins-little-known-passion-project-a-refrigerator/.

Clymer Marlay Noble Jr.: "Guide for the Uncertain: Advice for Life by Einstein to a Novice Chemist. Princeton, 1946." Winner's Auctions. Accessed August 22, 2022. https://winners-auctions.com/en/items/guide-for-the-uncertain-advice-for-life-by-einstein-to-a-novice-chemist-princeton-1946/.

letters to Einstein seeking advice and counsel: Ziv, Stav. "Albert Einstein Letters on Science, Politics and God Could Be Yours." *Newsweek*, June 13, 2017. https://www.newsweek.com/albert-einstein-letters-science-politics-god-auction-625290.

sold at auction: "Einstein Letters on God, Israel and Physics Fetch $210,000." *Phys.org*, June 20, 2017. https://phys.org/news/2017-06-einstein-letters-god-israel-physics.html.

WITH A NAME LIKE EINSTEIN

"It was the most hilarious routine": "Friars Club—Harry Einstein (Aka Parkyakarkus)." YouTube, November 15, 2015. https://www.youtube.com/watch?v=KuVA-IbuQiU.

"the real Albert Einstein changed his name": Moran, Edward. "Rob Reiner Plans New Documentary on Comedian Albert Brooks." *Cinema Daily US*, May 18, 2022. https://cinemadailyus.com/news/rob-reiner-plans-new-documentary-on-comedian-albert-brooks/.

"We have a history of building and buying places": Jeff Einstein, interview with the author, November 19, 2021.

"the Mick Jagger of digital media": Mahler, Jonathan. "Commute to Nowhere." *New York Times*, April 13, 2003. https://www.nytimes.com/2003/04/13/magazine/commute-to-nowhere.html.

"A choice between Schwartz and Einstein": Mara Einstein, interview with the author, November 11, 2021.

"I had an inferiority complex": Cayla Einstein, interview with the author, November 10, 2021.

"I embrace it": Stephen Einstein, interview with the author, June 4, 2021.

"My children are considered ethnically diverse": Rebecca Einstein, interview with the author, November 10, 2021.

"Back in my day": Laurie Einstein Koszuta, interview with the author, June 7, 2021.

played in a band with Albert Einstein: "George S. Einstein Obituary (1919—2011) Ocala Star-Banner." Legacy.com. Accessed August 22, 2022. https://www.legacy.com/us/obituaries/ocala/name/george-einstein-obituary?id=20670044.

"my theory of relatives": Daniel Einstein, interview with the author, June 17, 2021.

"It's a double-edged sword": Ted Einstein, interview with the author, June 22, 2021.

"My students are always really disappointed": Sarah Einstein, interview with the author, June 11, 2021.

"I do feel kinship": Einstein, Sarah. *Ulm: Part 1*. Writing Family Histories, April 25, 2022. https://writingfamilyhistories.substack.com/p/ulm-part-1.

at least half of the regiment were German immigrants: Sarna, Jonathan D., and Benjamin Shapell. *Lincoln and the Jews: A History*. United States: St. Martin's Publishing Group, 2015, 88.

"The only thing I inherited": Karen Cortell Reisman, interview with the author, March 7, 2019.

EINSTEIN AND POP CULTURE

"Lawrence of Arabia meets Bugs Bunny": Forsberg, Myra. "The Name's Serious. Yahoo Serious." *New York Times*, July 30, 1989. https://www.nytimes.com/1989/07/30/movies/film-the-name-s-serious-yahoo-serious.html.

"dumber than a bowling ball": Kempley, Rita. "'Young Einstein.'" *Washington Post*, August 4, 1989. https://www.washingtonpost.com/wp-srv/style/longterm/movies/videos/youngeinsteinpgkempley_a09fd0.htm.

Lisa Simpson saw a sign: "I Know Those Words, but That Sign Makes No

Sense." YouTube, March 22, 2014. https://www.youtube.com/watch ?v=xq0XNILIYTw.

A reporter for *The Guardian***:** Buckmaster, Luke. "Forget the Pyramids, the Greatest Mystery of Our Time Is What Happened to Yahoo Serious | Luke Buckmaster." *The Guardian*, June 3, 2019. https://www.theguardian.com/tv -and-radio/2019/jun/03/forget-the-pyramids-the-greatest-mystery-of-our -time-is-what-happened-to-yahoo-serious.

"G'day, Benyamin": Yahoo Serious, interview with the author, July 13, 2022.

"belief in the goodness of scientific inquiry": Grobar, Matt. "'Genius' Star Geoffrey Rush on 'Humanizing' Einstein, an Iconic Figure We Only Thought We Knew." *Deadline*, August 10, 2017. https://deadline.com/2017/08/genius -geoffrey-rush-ron-howard-emmys-interview-news-1202141175/.

"I'd read other feature scripts": Radish, Christina. "Ron Howard on Genius, Albert Einstein, and the Dark Tower." *Collider*, April 24, 2017. https://collider.com/ron -howard-genius-interview/.

"Having people like Einstein in the forefront of popular culture": Mosher, Dave. "'Genius' Director Ron Howard Reveals Why He's on a Mission to Turn Scientists into Celebrities." *Business Insider*, April 29, 2017. https://www .businessinsider.com/ron-howard-genius-einstein-american-science-2017-4.

"this guy isn't quite what he seems": Grazer, Brian, and Charles Fishman. *A Curious Mind: The Secret to a Bigger Life*. United Kingdom: Simon & Schuster, 2016, 122.

"the intersection of science": Lauren Gussis, interview with the author, May 18, 2021.

"Everything about him is electric": Photograph of a painting of Albert Einstein, inscribed and signed by Einstein. Accessed August 22, 2022. https://www .pbagalleries.com/view-auctions/catalog/id/291/lot/88553/Photograph -of-a-painting-of-Albert-Einstein-inscribed-and-signed-by-Einstein.

In a rap song: Platon, Adelle. "Kanye Brags He's the Next Einstein, Talks Twitter Rants on New Song." *Billboard*, February 24, 2016. https://www.billboard.com/music /rb-hip-hop/kanye-west-next-einstein-twitter-debut-new-song-6889411/.

An entire episode of: Allen, Ben. "Netflix's Russian Doll Timelines Explained." *Radio Times*, February 8, 2019. https://www.radiotimes.com/tv/sci-fi/russian -doll-netflix-explained-orange-speech-relativity-timelines-loops-what-is -happening/.

"solved the coat problem for years": "Thomas Venning: 5 Minutes with Einstein's Leather Jacket." *Christies*, July 11, 2016. https://www.christies.com/features /Thomas-Venning-Books-Specialist-examines-Einsteins-leather-jacket-7510–1.aspx.

The winning bid: Mantor, Cassidy. "Levi's Wins Einstein's Jacket at Christie's Auction." *FashionNetwork.com*, July 14, 2016. https://ww.fashionnetwork.com /news/levi-s-wins-einstein-s-jacket-at-christie-s-auction,713886.html.

"This jacket is just one more example": Ciment, Shoshy. "Levi's Made an Exact Replica of Albert Einstein's Iconic Leather Jacket and Sold It." *Business Insider,* December 5, 2019. https://www.businessinsider.co.za/levis-sold-albert -einstein-iconic-leather-jacket-for-1200-2019-12.

the death of Robin Williams: Hoffner, Cynthia A., and Elizabeth L. Cohen. "Mental Health-Related Outcomes of Robin Williams' Death: The Role of Parasocial Relations and Media Exposure in Stigma, Help-Seeking, and Outreach." *Health Communication* 33, no. 12 (2017): 1573–82. https://doi.org /10.1080/10410236.2017.1384348.

"the paradox of ambient awareness": Thompson, Clive. "Brave New World of Digital Intimacy." *New York Times*, September 5, 2008. https://www.nytimes .com/2008/09/07/magazine/07awareness-t.html.

"semiotic systems": Turner, Graeme. "Understanding Celebrity," 2004. https://doi .org/10.4135/9781446279953.

"because everybody loves him": Rebecca Watson, interview with the author, October, 27, 2017.

cakes in the shape of the Incredible Hulk and ALF: Inci Orfanli Erol, interview with the author, November 1, 2016.

a cop-turned-cake decorator: Dawn Butler, interview with the author, April 21, 2016.

"I was crucified twice nightly for a month": Jake Broder and David Ellenstein, interview with the author, May 3, 2021.

"everybody wants Einstein": Duffy Hudson, interview with the author, March 1, 2016.

EINSTEIN IN THE AGE OF FAKE NEWS

It was election night: "670: Beware the Jabberwock." *This American Life*, February 1, 2021. https://www.thisamericanlife.org/670/transcript.

fore more than a year: "Connecticut State Police Sandy Hook Elementary School Shooting Report." Accessed August 22, 2022. https://cspsandyhookreport .ct.gov/.

Lenny told me this: Lenny Pozner, interview with the author, June 16 and 17, 2022.

using Noah's photo without his permission: Sway. "The Sandy Hook Father Who Refused to Let Alex Jones Win." *New York Times,* February 10, 2022. https://www.nytimes.com/2022/02/10/opinion/sway-kara-swisher-leonard -pozner.html?showTranscript=1.

Hackers in 2013: Domm, Patti. "Markets Sink Briefly on Fake AP Terror Tweet." *CNBC,* April 24, 2013. https://www.cnbc.com/id/100646197.

A 2019 survey by the Pew Research Center: Mitchell, Amy, Jeffrey Gottfried, Galen Stocking, Mason Walker, and Sophia Fedeli. "Many Americans Say Made-up News Is a Critical Problem That Needs to Be Fixed." *Pew Research Center,* November 12, 2021. https://www.pewresearch.org/journalism/2019/06/05 /many-americans-say-made-up-news-is-a-critical-problem-that-needs-to-be -fixed/.

Fake news did not begin: Bowman, N.D., and E. L. Cohen. "Technologies of mass deception? War of the Worlds, Twitter, and a history of fake and misleading news in the United States." In E. Downs (Ed.), *The Dark Side of Media & Technology* (pp. 25 -36). New York: Peter Lang, 2019. doi:10.3726/b14959.

He copied Pelham's painting: "Boston Massacre and Propaganda: Changing Depictions of Crispus Attucks." Museum of the American Revolution. Accessed August 23, 2022. https://www.amrevmuseum.org/boston-massacre -and-propaganda-changing-depictions-of-crispus-attucks.

he printed a hoax supplement: "Founders Online: 'Supplement to the Boston Independent Chronicle.'" National Archives and Records Administration. Accessed August 23, 2022. https://founders.archives.gov/documents/Franklin /01-37-02-0132.

"defenseless farmers": "Founders Online: From Benjamin Franklin to Dumas, 3 May 1782." National Archives and Records Administration. Accessed August 23, 2022. https://founders.archives.gov/documents/Franklin/01-37-02-0180.

"His thoughtful friends": Stanley, Matthew. *Einstein's War: How Relativity Triumphed Amid the Vicious Nationalism of World War I.* United Kingdom: Dutton, 2019, 97.

"My sense of it is": Matthew Stanley, interview with the author, June 10, 2022.

"Blind obedience to authority": Calaprice, Alice. *The Ultimate Quotable Einstein.* United States: Princeton University Press, 2019, 161.

eight hundred million new tweets: "How Many Tweets per Day 2022 (Number of Tweets per Day)." *Renolon,* August 8, 2022. https://www.renolon.com /number-of-tweets-per-day/.

three billion active users per month: "Meta Reports Second Quarter 2022 Results." *Meta.* Accessed August 23, 2022. https://investor.fb.com/investor

-news/press-release-details/2022/Meta-Reports-Second-Quarter-2022
-Results/default.aspx.

five hundred hours of video: "More than 500 Hours of Content Are Now Being Uploaded to YouTube Every Minute." *Tubefilter*, May 7, 2019. https://www.tubefilter.com/2019/05/07/number-hours-video-uploaded-to-youtube-per-minute/.

brought in more than $50 million annually: Williamson, Elizabeth. "Alex Jones and Donald Trump: A Fateful Alliance Draws Scrutiny." *New York Times*, March 7, 2022. https://www.nytimes.com/2022/03/07/us/politics/alex-jones-jan-6-trump.html.

filed for bankruptcy: Spector, Mike, and Dietrich Knauth. "Conspiracy Website *InfoWars* Parent Files for Bankruptcy." *Reuters*, July 30, 2022. https://www.reuters.com/business/media-telecom/infowars-parent-files-bankruptcy-2022-07-29/.

A Georgia poll worker: Shivaram, Deepa. "Shaye Moss Staffed an Election Office in Georgia. Then She Was Targeted by Trump." NPR, June 22, 2022. https://www.npr.org/2022/06/22/1106459556/shaye-moss-staffed-an-election-office-in-georgia-then-she-was-targeted-by-trump.

it's an echo chamber: Pariser, Eli. *The Filter Bubble: How the New Personalized Web Is Changing What We Read and How We Think.* New York: Penguin Books, 2012.

"creates the illusion of knowing": Postman, Neil. *Amusing Ourselves to Death: Public Discourse in the Age of Show Business.* United Kingdom: Penguin Publishing Group, 2006, 75.

"In the pre-internet era": Stengel, Richard. "Misdirection, Fake News and Lies: The Best Books to Read on Disinformation." *New York Times*, June 9, 2022. https://www.nytimes.com/2022/06/09/books/books-disinformation-fake-news.html.

"knowledge began freely replicating": Fallows, James. "The 50 Greatest Breakthroughs since the Wheel." *The Atlantic*, August 24, 2015. https://www.theatlantic.com/magazine/archive/2013/11/innovations-list/309536/.

A study out of Princeton University: Travers, Mark. "Facebook Spreads Fake News Faster than Any Other Social Website, According to New Research." *Forbes*, December 10, 2021. https://www.forbes.com/sites/traversmark/2020/03/21/facebook-spreads-fake-news-faster-than-any-other-social-website-according-to-new-research/?sh=54f6a3cc6e1a.

An analysis from Buzzfeed: Silverman, Craig. "This Analysis Shows How Viral Fake Election News Stories Outperformed Real News on Facebook." *BuzzFeed News*, November 16, 2016. https://www.buzzfeednews.com/article/craigsilverman/viral-fake-election-news-outperformed-real-news-on-facebook.

"The first casualty of war is truth": "Pivot: Truth Social Sinks, More Russia Fallout for Big Tech, and Guest Elizabeth Williamson on Apple Podcasts." *Pivot/Vox Media*. Apple Podcasts, March 8, 2022. https://podcasts.apple.com/us/podcast /truth-social-sinks-more-russia-fallout-for-big-tech/id1073226719?i =1000553288572.

William Frauenglass: Jerome, Fred. *The Einstein File: J. Edgar Hoover's Secret War against the World's Most Famous Scientist*. United States: St. Martin's Press, 2003, 238–239.

"enemy of America": Jerome, Fred. *The Einstein File: J. Edgar Hoover's Secret War against the World's Most Famous Scientist*. United States: St. Martin's Press, 2003, 240.

"I suggest he move to Russia": "Einstein, Plumbers, and McCarthyism." Institute for Advanced Study, 2017. https://www.ias.edu/ideas/2017/einstein -mccarthyism.

the Woman Patriot Corporation: Jerome, Fred. *The Einstein File: J. Edgar Hoover's Secret War against the World's Most Famous Scientist*. United States: St. Martin's Press, 2003, 6.

"The annihilation of any life": "February 12: Einstein Opposes the H-Bomb." Jewish Currents. Accessed August 23, 2022. https://jewishcurrents.org/february -12-einstein-opposes-the-h-bomb.

The very next morning: Jerome, Fred. *The Einstein File: J. Edgar Hoover's Secret War against the World's Most Famous Scientist*. United States: St. Martin's Press, 2003, 156.

"If it got out prematurely": Jerome, Fred. *The Einstein File: J. Edgar Hoover's Secret War against the World's Most Famous Scientist*. United States: St. Martin's Press, 2003, XXI.

that bulged to 1,427 pages: Waldrop, Mitch. "Why the FBI Kept a 1,400-Page File on Einstein." *National Geographic*, May 4, 2021. https://www.nationalgeographic .com/pages/article/science-march-einstein-fbi-genius-science.

which the FBI finally made available online: "Albert Einstein." FBI, December 6, 2010. https://vault.fbi.gov/Albert%20Einstein.

"He was largely indifferent": Rao, Rahul. "Why Was the FBI Obsessed with Albert Einstein?" *Inverse*, March 22, 2022. https://www.inverse.com/science/fbi -obsessed-with-einstein.

"flourishes in times of uncertainty": Stengel, Richard. "Misdirection, Fake News and Lies: The Best Books to Read on Disinformation." *New York Times*, June 9, 2022. https://www.nytimes.com/2022/06/09/books/books-disinformation -fake-news.html.

by the polling firm Ipsos: "Why Misinformation Threatens the Future of Science." *Axios.* Accessed August 23, 2022. https://www.axios.com/sponsored/content -item/3m-why-misinformation-threatens-the-future-of-science.

Twitter tested a feature: Sparks, Hannah. "New Twitter Feature Challenges Users to Read Articles before Sharing." *New York Post,* June 11, 2020. https:// nypost.com/2020/06/11/twitter-feature-asks-users-if-theyve-read-content -before-sharing/.

"The problem begins": Calaprice, Alice. *The Ultimate Quotable Einstein.* United States: Princeton University Press, 2019, 328.

"a powerful and universal heuristic": Hoogeveen, Suzanne, Julia M. Haaf, Joseph A. Bulbulia, Robert M. Ross, Ryan McKay, Sacha Altay, Theiss Bendixen, et al. "The Einstein Effect Provides Global Evidence for Scientific Source Credibility Effects and the Influence of Religiosity." *Nature Human Behaviour* 6, no. 4 (2022): 523–35. https://doi.org/10.1038/s41562-021-01273-8.

PRESERVING EINSTEIN

They loaded up all his papers: Chaya Becker, interview with the author, June 12, 2022.

"the mass frenzy": Isaacson, Walter. *Einstein: His Life and Universe.* United States: Simon & Schuster, 2017, 289.

"lent the prestige mondial": *Albert Einstein, Historical and Cultural Perspectives: The Centennial Symposium in Jerusalem.* United States: Princeton University Press, 2014, 283.

"Palestine is not primarily": "Prof. Einstein Discusses Judaism." *Jewish Telegraphic Agency,* September 30, 1934. https://www.jta.org/archive/prof -einstein-discusses-judaism.

"There is only one man": *Albert Einstein, Historical and Cultural Perspectives: The Centennial Symposium in Jerusalem.* United States: Princeton University Press, 2014, 293–296.

"Einstein was not wearing socks": Isaacson, Walter. *Einstein: His Life and Universe.* United States: Simon & Schuster, 2017, 523.

decided to bequeath: "Hebrew University v. General Motors LLC." Casetext, October 15, 2012. https://casetext.com/case/hebrew-univ-of-jerusalem-v-gen -motors-llc.

sixty million people: Isaacson, Walter. *Einstein: His Life and Universe.* United States: Simon & Schuster, 2017, 541.

the world's leading Einstein expert: Roni Grosz, interview with the author, April 2, 2018.

have Einstein's fingerprints on them: Neil McManus, interview with the author, March 6, 2019.

110 new pages: Ratner, Paul. "The Missing 'Puzzle' Page of Einstein's Unified Theory of Everything Found." *Big Think*, March 10, 2019. https://bigthink .com/hard-science/the-missing-puzzle-page-of-einsteins-unified-theory-of -everything-found/.

"The memory of Einstein": Hanoch Gutfreund, interview with the author, April 2, 2018.

perhaps how Gutenberg did it: Ido Agassi, interview with the author, March 11, 2019.

"was not particularly concerned": Matthew Stanley, interview with the author, June 10, 2022.

EINSTEIN AND THE NEXT GENERATION

while in the car: Shuki Levy, interview with the author, February 17, 2017.

the world's leading collector of Andy Warhol paintings: "Jewish Art Collector Jose Mugrabi Awarded Honorary Doctorate by Hebrew U." *Jerusalem Post*, June 14, 2022. https://www.jpost.com/j-spot/article-709419.

"a collector of characters": Doron Ofir, interview with the author, November 6, 2019.

"He respected children": *Dear Professor Einstein: Albert Einstein's Letters to and from Children.* United States: Prometheus, 2010, 11.

2015 in California: "Local School Sets Record for Albert Einstein Look-Alikes." Today at Berkeley Lab, November 17, 2015. https://today.lbl.gov/2015/11/17 /local-school-sets-record-for-albert-einstein-look-alikes/.

elementary school in Toronto: Zax, Talya. "A Bunch of Canadian Albert Einstein Look-Alikes Just Broke a Guinness World Record." *The Forward*, March 30, 2017. https://forward.com/culture/367586/a-bunch-of-canadian-albert-einstein -look-alikes-just-broke-a-guinness-world/.

541 students: "Largest Gathering of People Dressed as Albert Einstein." Guinness World Records. Accessed August 23, 2022. https://www .guinnessworldrecords.com/world-records/100331-largest-gathering-of-people -dressed-as-albert-einstein.

"As the alcohol level in our blood rose": Yonatan Winetraub, interview with the author, February 19, 2019.

offered $20 million: "The New Space Race." XPRIZE. Accessed August 23, 2022. https://www.xprize.org/prizes/google-lunar.

Google withdrew the prize money: Chang, Kenneth. "The Google Lunar X Prize's Race to the Moon Is Over. Nobody Won." *New York Times*, January 23, 2018. https://www.nytimes.com/2018/01/23/science/google-lunar-x-prize-moon.html.

raised through private donations: i24NEWS. "Israel's SpaceIL Raises $70M for Another Shot at Moon Landing." *I24NEWS*, July 11, 2021. https://www.i24news.tv/en/news/israel/technology-science/1626002095-israel-s-spaceil-raises-70m-for-another-shot-at-landing-on-moon.

"This is an investment in science": Sylvan Adams, interview with the author, March 12, 2019.

"Condolences to the Beresheet lander": Steinbuch, Yaron. "Buzz Aldrin Offers Condolences after Israel's Failed Moon Landing." *New York Post*, April 12, 2019. https://nypost.com/2019/04/12/buzz-aldrin-offers-condolences-after-israels-failed-moon-landing/.

"It's particularly meaningful": Atara Solow, interview with the author, April 11, 2019.

344 things to be exact: Shapiro, Ari. "The James Webb Telescope Had 344 'Single Point Failures' before Launch. Then, Success." *OPB*, July 17, 2022. https://www.opb.org/article/2022/07/17/james-webb-telescope-had-344-single-point-failures-before-launch-then-success/.

"The luminous cores of galaxies": Avi Loeb, interview with the author, July 17, 2022.

"I headed to the hospital first": Ronk, Liz. "Genius in the Details: Revisiting an Iconic Photo of Albert Einstein's Office." *Time*, February 12, 2014. https://time.com/3879070/albert-einstein-last-photo-taken-of-his-princeton-office/feed/.

"embarking on this cosmic quest": Kaku, Michio. "In a Parallel Universe, Another You." *New York Times*, June 20, 2022. https://www.nytimes.com/2022/06/20/special-series/michio-kaku-multiverse-reality.html.

"strums the deepest chords": Kean, Sam. "The Face of Science." The National Endowment for the Humanities, May 2014. https://www.neh.gov/humanities/2014/mayjune/feature/the-face-science.

INDEX

ABOUT THE AUTHOR

Photo credit: Shoshi Benstein

Benyamin Cohen manages the official social media accounts of Albert Einstein. He is the News Director of the Forward and was the founding editor of both *Jewsweek* and *American Jewish Life* magazine. Cohen is also the author of *My Jesus Year: A Rabbi's Son Wanders the Bible Belt in Search of His Own Faith*, named one of the best books of the year by *Publishers Weekly* and for which he received the Georgia Author of the Year award. He is based in Morgantown, West Virginia, where he lives with his wife, three dogs, and a flock of chickens known as the Co-Hens.

For more information: www.benyamincohen.com

Follow @alberteinstein online at
Facebook, Instagram, and Twitter.